THE MITCHELL BEAZLEY POCKET GUIDE TO THE
SEASHORES
of Britain and Northern Europe

John M. Baxter

MITCHELL BEAZLEY

The Mitchell Beazley Pocket Guide to the Seashores of
Britain and Northern Europe

Edited and designed by Mitchell Beazley International Ltd.,
Artists House, 14-15 Manette Street, London W1V 5LB

Editor	Simon Ryder
Art Editor	Christopher Howson
Production	Barbara Hind
Executive Editor	Robin Rees

Artwork by Ron Hayward, John Hutchinson and
George Thompson
Indexed by Annette Musker

British Library Cataloguing in Publication Data for this book are
available from the British Library

ISBN 0 85533 769 9

Typeset in Bembo by Servis Filmsetting Ltd., Manchester
Reproduced by Mandarin Offset, Hong Kong
Produced by Mandarin Offset
Printed and bound in Hong Kong

CONTENTS

Using This Book

The Seashore Environment (pages 8-21) introduces the physical features that shape the seashore and the variety of life that can be found there.

Exploring the Seashore (pages 22-41) describes the range of seashore habitats, the species they contain, and how to explore them.

Identifying the Species (pages 42-138) is a field guide to around 400 of the animals and plants that are commonly found on the coasts of Britain and Northern Europe. The following habitat symbols are included to help with identification.

🌿	On rock	⊜	On sand or mud
🌺	On a plant	⬙	On an animal
⁘	In gravel or amongst pebbles	🐚	Under an overhang
〰	In a rockpool	🌀	Under seaweed
▤	In an estuary	🔆	In a saltmarsh
⬤	In sand or mud	🐚	On cliffs

Species names

In the introductory sections, the common name of each species is used wherever possible. But this can be unreliable as it often varies from place to place. Not only may a single species have a number of different common names (one particular red seaweed has three names: Dulse, Sheep's Weed and Dillisk), but one common name may refer to two different species (in the Orkney Islands the edible winkle is known as a whelk, whereas elsewhere the name whelk refers to a completely different mollusc). In addition, many of the less common species do not have common names.

In contrast, the scientific name of each species is determined by strict international rules. It consists of two parts; first the genus name, second the species name. For example, the scientific name of the red seaweed mentioned above is *Palmaria palmata*. In the identification section both the scientific name and common name(s) are given.

When referring to all the species within a particular genus, the species names are dropped and replaced by *spp*. So, all the seaweeds in the genus *Palmaria* are referred to as *Palmaria spp*.

Introduction

The attraction of the seashore lies in its sights and in its sounds, as well as in the pleasurable sensations of the sand and the sea. It is a busy environment, with ever changing and frequently unpredictable conditions. But it is precisely these dynamic conditions – tides, waves, exposure, shelter, salinity, temperature – and the habitats that result that make the seashore so fascinating. To find such a wide range of habitats inland you would have to travel far greater distances, and to gain the same overview of the mosaic of habitats you would have to take to the air.

The seashore weaves a complex web of intimate relationships between plants and animals and their habitats. Perhaps of all environments, it is the one where the visitor needs to be a true naturalist, part botanist, part zoologist and part ecologist. Identifying the plants and animals is very satisfying in itself, but if you have some understanding of the natural processes that have placed them there then you will begin to appreciate just what an exceptional environment the seashore is.

This book is intended very much as a companion and a field guide, allowing identification as well as offering a fuller understanding of the whole environment. It is a departure from other guides to the seashore in that it includes, in addition to the animals and plants of the shore itself, the more common coastal flowering plants, the birds of the shore and coastal waters, and some mammals. This is in order to present a selection of species that matches the range you are likely to encounter on a visit. It is inevitable, and it must be stressed, that animals and plants will be found that are not described in this book. To present a description of everything that could be found would be impracticable, and the result would certainly not be pocket-sized. It is also intended that all the species described should be identifiable with no more than a hand lens. For those who wish to know more, some relevant texts are recommended at the back of the book.

There are differences between individuals in any species of plant or animal, so the descriptions and illustrations in the identification section should be regarded as guides to the normal form of the species. The size limits are realistic estimates, though larger individuals may be found, and the colours are by no means always definitive. However, in identification terms, the habitat does have a fairly rigorous controlling influence on what occurs where – for example, you would never find a limpet living in mud or sand. The use of habitat symbols in the identification section reflects this.

Another intriguing feature of the seashore is the seasonal variation in species. For example, seals beaching to give birth to their pups during the summer, or the winter migration of wading birds to food-rich mudflats. Less obvious, but just as interesting, is the migration of certain invertebrates like the sea lemon (a species of sea slug), which will be much more common on the shore during the summer, when it moves up from deeper water to spawn.

Every time you go to the seashore it is possible to discover something new, and understanding it is like trying to solve a self-perpetuating puzzle, since answering one question merely raises many more. I hope that this book will provide some enlightenment and increase your enjoyment of the seashore.

ICELAND

NORWEGIAN SEA

ATLANTIC OCEAN

➂⑤

14°/4°

15°/8°

14°/7°

NORTH SEA

NORWAY

14°/10°

➂⑤

N. IRELAND

➂④

IRELAND

GREAT BRITAIN

14°/5°

➂⑤

➁⑧

DENMARK

17°/4°

➂④

15°/9°

14°/8°

ENGLISH ➂⑤ CHANNEL

BELGIUM

HOLLAND

WEST GERMANY

EAS

17°/8°

FRANCE

BAY OF BISCAY

SPAIN

KEY

0-200 metres

200-2,000 metres

2,000-4,000 metres

4,000+ metres

➂⑤ Circled figures show the average salinity of the seawater in parts per thousand.
15°/9° Summer and winter average surface temperatures are given in degrees centigrade (summer°/winter°)
➡ Arrows indicate the direction and relative strengths of currents

GULF
OF
BOTHNIA
⑥ 13°/0°

SWEDEN

FINLAND

17°/3°

BALTIC SEA

⑩

RUSSIA
(USSR)

POLAND

ERMANY

From the influence of the Gulf
Stream to the Gulf of Bothnia, the
environmental conditions in the
seas around Britain and Northern
Europe vary widely in temperature,
salinity and tides. Water from the
Gulf Stream is carried across the
Atlantic Ocean by the Atlantic
Current and keeps the west coasts
of Ireland and Scotland warm in
summer and mild in winter. The
shores bordering the North Sea are
washed by colder waters from the
north, but here the winter
temperature is moderated by the
high salinity. The waters of the
Baltic Sea are the coldest in this
region. They have the lowest
salinity, with ice forming in winter
in the Gulf of Bothnia. There is
also a large variation in tidal range;
with the Baltic Sea having virtually
no tides, while the maximum
spring tides elsewhere vary from
under one metre, on the northeast
coast of Ireland, to over ten metres
in the Bristol Channel, with an
average of three to four metres. It
is this outstanding diversity of
environmental conditions,
combined with the rich larvae-
bearing currents, that makes the
seashores of Britain and Northern
Europe so varied.

TIDES

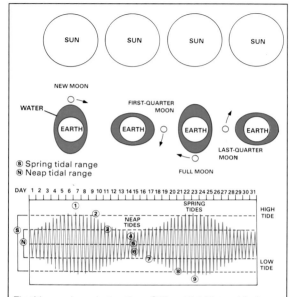

The tides are dependent on the relative positions of the Sun, Moon and Earth. The Moon's gravitational pull is strongest on the water closest to it, producing a bulge in the Earth's oceans. Its pull on the water on the other side of the Earth is weaker than on the Earth itself, so this water is effectively 'left behind', resulting in a second bulge.

① Extreme High Water of Spring tides (EHWS)

② Mean High Water of Spring tides (MHWS)
③ Mean High Water of Neap tides (MHWN)
④ Lowest High Water of Neap tides (LHWN)
⑤ Mean Tide Level (MTL)
⑥ Highest Low Water of Neap tides (HLWN)
⑦ Mean Low Water of Neap tides (MLWN)
⑧ Mean Low Water of Spring tides (MLWS)
⑨ Extreme Low Water of Spring tides (ELWS)

The rise and fall of the tides, as they cover and uncover the shore, make the seashore both a complex and demanding place for the animals and plants that live there, and a fascinating place to explore. The tides are the main factor controlling life on the shore, and it is useful to understand how they work and the effect that weather has on them to get the most out of any visit. Also, by knowing the state of the tide before setting-out you can choose the best time to visit.

Tides are produced by the gravitational pull of the moon, and to a lesser extent the sun, on the oceans. Because the moon is so much nearer to the earth than the sun it has the greater effect, despite the fact that the sun is so much larger. These gravitational forces combine to produce three tidal cycles: the regular, twice daily, coming-in (flooding) and going-out (ebbing), which depends on the moon's orbit around the earth; the monthly cycle of spring and neap tides, which coincides with the lunar cycle; and the annual cycle, which is a result of the earth orbiting the sun. The daily cycle is modified by the monthly one, which in turn is modified by the annual cycle.

Although the tides appear to move horizontally, up and down the shore, they are really the result of a vertical rising and falling of the sea. The extent of this vertical movement, known as the tidal range, varies both in time and from place to place. The tides with the greatest range occur just after the new and full moons and are known as spring tides (which has nothing to do with that season of the year). At this time the gravitational pull of the moon and sun are in roughly the same direction, combining to produce large tides. The largest spring tides of all occur in late March and September each year at the time of the solar equinoxes, when the moon and sun are exactly aligned. The smallest tides in each monthly cycle are known as the neap tides and occur around the time of the first- and last-quarter moons, when the gravitational pull of the moon is at right angles to that of the sun.

Spring and neap tidal ranges vary greatly from one point on the coast to another. The northeast coast of Ireland has an average spring tidal range of one metre and a neap tidal range of only half a metre, compared with twelve-and-a-half metres and six-and-a-half metres respectively for the southwest of England and parts of the Channel coast.

The time of day at which low water spring tide occurs is fairly constant at any point on the coast. Where it occurs early in the morning (and in the evening) a greater diversity of animals and plants are found in the lowest region of the shore, compared with shores where it occurs around the middle of the day, since more sensitive species cannot cope with being exposed during the warmest part of the day.

Tide tables give predictions of the tide times and ranges assuming 'average' weather conditions. The weather can affect both the size and the time of the tides. Long periods of strong winds tend to 'pile up' the sea in the direction towards which they are blowing. An onshore wind therefore has the effect of pushing both high and low water higher up the shore; an offshore wind has the reverse effect. An area of extreme high or low pressure can also affect the size of a tide. A low pressure system tends to suck up the seawater beneath it, whilst high pressure pushes it down. The height of a tide may be raised or lowered by as much as 50 centimetres. Consequently, a high pressure system occurring at the same time as large spring tides can result in areas of the shore being exposed which are normally covered at all times.

Whilst the tides are the most obvious factor influencing the distribution of animals and plants on the shore, other forms of water movement are also important. For example, the amount of wave action on the shore affects the distribution of species (see pages 10-11). A more subtle factor is the presence of ocean currents which, apart from influencing the temperature of the water (as with the Gulf Stream warming the west coasts of Ireland and Scotland), transport millions of larvae and spores around the seas (see pages 16-17).

EXPOSURE
AND ZONATION

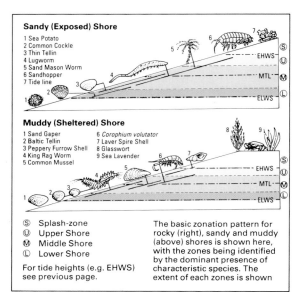

Sandy (Exposed) Shore

1 Sea Potato
2 Common Cockle
3 Thin Tellin
4 Lugworm
5 Sand Mason Worm
6 Sandhopper
7 Tide line

Ⓢ
Ⓤ   EHWS
Ⓜ   MTL
Ⓛ   ELWS

Muddy (Sheltered) Shore

1 Sand Gaper   6 *Corophium volutator*
2 Baltic Tellin   7 Laver Spire Shell
3 Peppery Furrow Shell 8 Glasswort
4 King Rag Worm   9 Sea Lavender
5 Common Mussel

Ⓢ
Ⓤ   EHWS
Ⓜ   MTL
Ⓛ   ELWS

Ⓢ Splash-zone
Ⓤ Upper Shore
Ⓜ Middle Shore
Ⓛ Lower Shore

For tide heights (e.g. EHWS)
see previous page.

The basic zonation pattern for
rocky (right), sandy and muddy
(above) shores is shown here,
with the zones being identified
by the dominant presence of
characteristic species. The
extent of each zones is shown

A cliff face that is constantly pounded by large waves is an extreme
example of exposure, and only the hardiest species can withstand
such a battering. At the other end of the scale, a sheltered inlet
exposed only to the gentlest of waves can generally support a much
greater diversity of animal and plant life. The degree of exposure
depends on the the amount and type of wave action; not only the
frequency and size of the waves, but also the shape and location of
the shore. A shore with a gentle slope and broken profile will be less
exposed than one with a relatively steep, unbroken slope, because it
absorbs the energy of the waves over a greater area.

Zonation is a way of grouping species according to the tidal level
on the shore at which they are commonly found. The location of
animals and plants reflects their ability to tolerate air (sun, rain and
wind), or immersion in water, for varying periods of time. A
simple appreciation of exposure and zonation will allow you to
interpret the relationships between the physical and biological
features that make up the seashore environment.

It is possible to assess the amount of exposure and the tidal level
of any point on the shore by looking at the species present, some of
which can be used as exposure indicators. For a particular niche
there is often one species that will occupy that niche on a sheltered
shore and a similar, but different species that can be found in the
equivalent niche on an exposed shore – for example, the two
brown seaweeds, dabberlocks and sugar kelp, are found in
equivalent niches on exposed and sheltered shores respectively. To
make a biological assessment of the tidal level you have to look at

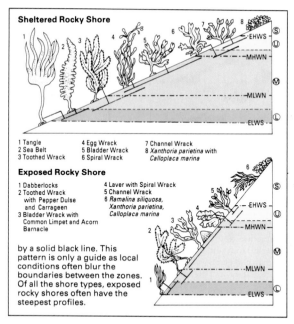

Sheltered Rocky Shore

1 Tangle
2 Sea Belt
3 Toothed Wrack
4 Egg Wrack
5 Bladder Wrack
6 Spiral Wrack
7 Channel Wrack
8 *Xanthoria parietina* with *Calloplaca marina*

Exposed Rocky Shore

1 Dabberlocks
2 Toothed Wrack with Pepper Dulse and Carrageen
3 Bladder Wrack with Common Limpet and Acorn Barnacle
4 Laver with Spiral Wrack
5 Channel Wrack
6 *Ramalina siliquosa*, *Xanthoria parietina*, *Calloplaca marina*

by a solid black line. This pattern is only a guide as local conditions often blur the boundaries between the zones. Of all the shore types, exposed rocky shores often have the steepest profiles.

the zonation pattern. There is a relatively narrow strip in the middle of the shore which is covered and uncovered by every tide. In contrast, the areas around the extreme high-water and extreme low-water marks are covered, or uncovered, relatively infrequently, dependant on the monthly cycle of spring and neap tides. How this is related to the distribution of species is shown above.

On rocky shores, which show the effects of exposure and zonation most clearly, there is a distinct succession of animals and plants from the top to the bottom of the shore (see page 27). No species spans the entire shore, but some have larger ranges than others. Mobile animals tend to have relatively wide ranges with ill-defined boundaries; their numbers gradually peter out towards the limits of their range, with the occasional outlying specimen extending the limit. In comparison, plants and sessile (stationary) animals have narrower ranges, with clear-cut boundaries. It is an interesting challenge when standing on a shore to look around you and, from the species that are closest to you, decide at what tidal level you are. With experience you can arrive at very accurate assessments.

Both the overall exposure rating of a shore and the basic pattern of zonation can be disrupted by localized conditions. An area may be influenced by a freshwater runoff, whilst elsewhere, increased shade and shelter may be afforded by a large crevice or a boulder. Such features can result in the presence of uncharacteristic species. If you are able to understand the reasons for such anomalies, then you are likely to appreciate and enjoy the seashore much more.

SHORE TYPES

Rocky and Shingle Shores

Sandy and Muddy Shores

There are four, fundamentally different types of shore: rocky, shingle, sandy and muddy. Each has its own typical community of animals and plants, which are best adapted to the conditions on that shore; but, each of the communities can be significantly modified in various ways by local conditions. Seldom will a rocky shore be found without the occasional pocket of sand or mud, or a sandy shore without a rock outcrop or a few boulders. The descriptions given below are, of necessity, very general, but they are made in an attempt to draw attention to the diversity of habitats.

Rocky shores

This type of shore varies considerably in its physical character, depending on the rock type (whether the rock is in the form of bedrock, boulders or pebbles) and the slope of the rock strata.

Bedrock shores range from near vertical cliffs, with a narrow intertidal region, to horizontal rock platforms, where the tidal range covers a large area. The animals and plants are visible and readily accessible on all but the steepest sloping shores and cliffs, making rocky shores particularly attractive to the visiting naturalist. Depending on the rock type, sloping and platform shores can provide, in addition to the flat surfaces, a multitude of crevices, cracks, overhangs and rockpools all worth exploring.

Stable boulder shores are common on more sheltered stretches of coastline. They are made up of boulders and pebbles of a wide range of sizes, embedded in sand or gravel. All the available surfaces are used, the undersides often sporting amazingly luxuriant and diverse growths of encrusting and mobile animals. It is also worth examining the wildlife in the sand and gravel in which the boulders are embedded.

Harbour walls, marina breakwaters, pier supports and other man-made structures are also interesting to explore. They can be considered as rocky shores, because the animals and plants do not differentiate between artificial and natural surfaces. Particularly on vertical walls, you will often find excellent examples of zonation.

Shingle shores

These shores could not be more different from rocky shores. Animal and plant life on a rocky shore is usually abundant and highly visible: shingle shores appear, and indeed are, deserted and barren. The pebbles making up the shore are continually rolled over each other by the action of the waves, and this prevents any life from becoming established. The gaps between the pebbles are too large to allow any water to collect when the tide ebbs, which means that even the lower, more stable layers are unsuitable for life.

Sandy shores

On first inspection, sandy shores do not appear to support much life. However, there are usually signs on the surface of the sand that show you where to look; often, there are large numbers of animals adapted to living in the sand at varying depths.

Usually gently sloping, sandy shores are made up of small particles of quartz. When the tide goes out, water is retained in the spaces between the particles, which lubricates the sand and creates a damp environment suitable for animals to inhabit. Even with the small sizes involved with sand particles, there are significant differences which affect the physical nature of the shore and so the biological nature as well. These differences are related to the degree of exposure of the shore to wave action; the greater the exposure the less likely that small sand grains will be deposited.

As you walk out over the sand, look at your feet. If you see a lighter coloured patch appear around each foot then you know you are on a beach of relatively coarse sand. Your weight has opened up the packing of the sand grains and there is not enough retained water to fill the enlarged spaces. As a result, there is increased friction between the particles and so greater resistance to movement through the sand. Try pushing your finger into the sand both next to your feet and as far away as you can reach. You will find that it is much more difficult near to where you are standing. Alternatively, your tread may make the surrounding area wetter and the sand become softer. This is indicative of fine sand grains and very important to many burrowing animals, because their own motion softens the sand making it easier for them to proceed.

Muddy shores

Unlike an inviting expanse of golden sand, a muddy shore never looks particularly appealing; you are faced with a dull grey or black mire, which is impossible to walk on without sinking and sliding. Do not be put off though; muddy shores have a great deal to offer, just be prepared to get dirty. Like sandy shores they are usually gently sloping or flat, hence the name mudflat.

Mudflats are made up of very fine particles of mineral and organic origin. The spaces between the particles are correspondingly small, which means that the mud does not dry out when the tide retreats. This ability to retain water, combined with the food potential of the organic matter, makes the mudflat a highly desirable habitat for some species of burrowing animals. These conditions, however, do not suit all animals, as the mud particles clog their delicate feeding and breathing mechanisms.

FLORA AND FAUNA

Purple Sunstar

Solaster endeca

KINGDOM
Animal

PHYLUM
Echinodermata

CLASS
Asteroidea

ORDER
Spinulosa

FAMILY
Solasteridae

GENUS
Solaster

SPECIES
endeca

Trying to relate all the seashore species is a complex task. Although a species' common name can be very descriptive, as with the Purple Sunstar, it's scientific name is part of a hierachical classification system which shows its relation to other species.

Apart from spectacular scenery and fresh air, the major attraction of the seashore is the superb array of animals and plants that live there. This diversity stems from the many different habitats found within the relatively short distance between the lowest low water mark and the upper limit of the splash-zone. This section is intended to give an overall view of that diversity.

Plants

There are three types of plant that you will come across on a visit to the coast: algae (seaweeds) which are more or less confined to the shore; lichens which inhabit both marine and terrestrial habitats; and flowering plants which occur mainly on the coastal fringe.

The seaweeds are usually separated into three groups by their colour; green, brown and red. This apparently obvious division looks like a useful first step in field identification; but it is not that easy since the question "What is brownish-red as opposed to reddish-brown?" often has to be answered. All seaweeds have a holdfast, which anchors them to the rocks, a stipe (stalk) and a frond; these three features are important for identification.

In the identification section of this book the typical or classic growth form is described for most seaweeds, but local conditions can alter this so take care when trying to identify a specimen. For example, in the genus *Fucus* there is a considerable degree of crossbreeding, which results in plants showing a combination of characters from the two parents.

Lichens are usually low growing or encrusting, branched and often brittle. Although a few species occur in the intertidal zone, they are most common, and more or less dominant, in the splash-zone, occurring in spectacularly colourful microforests.

Flowering plants are the group that people are most familiar with. They have adapted to many terrestrial and freshwater habitats, while only a very few have successfully made the transition to the marine environment; some of these are found in the genus *Zostera*, the eelgrasses. However, many flowering plants have developed a degree of tolerance to the marine environment; these are the plants of the saltmarshes, coastal fringes and cliffs.

Animals

Nowhere is the diversity of animal life better represented than on the shore. The complete range of complexity is present from the simplest sponges, through the many and the varied invertebrate phyla, to the fishes, birds and mammals.

Sponges are the simplest animals dealt with here, but specimens of a single species can vary in both form and colour, making them difficult to classify.

Sea anemones have a standard, instantly recognizable structure, but again colour variations make field identification hazardous. Their close relatives, the sea firs are mostly small, insignificant, colonial animals, which inhabit almost every available niche.

There is an enormous diversity of worm-like animals, including ribbon worms (nemertines), segmented worms (annelids) and cylindrical worms. Many of them are burrowers or tube dwellers.

Crustacea also have a relatively standardized design with segmented bodies and paired appendages. Many of the larger or more common species are easily identified in the field, such as the shrimps, crabs and lobsters, with a few exceptions; barnacles, for example, are crustacea despite their white calcareous shells.

Molluscs present a dazzling array of forms; the limpets and snails with their conical or coiled shells, the chitons with eight overlapping shell valves, the bivalves with a pair of hinged shells, the sea slugs which are the most delicately beautiful, brightly coloured members of the phylum and finally, the pinnacle of molluscan evolution, the squid and octopus – how different they appear from the limpet or the winkle.

A much less familiar group is the sea mats. These encrusting, colonial animals are found on most surfaces, including rock, shell and seaweed. There are a lot of species, all of which are very difficult to identify individually on the shore.

Echinoderms, in the shape of starfish, brittle stars, feather stars, sea urchins and sea cucumbers, have come to be regarded as an essential component of any seashore. They are a varied group of animals unique to the marine environment.

Unexpected as it may seem, the sea squirts are on the top of the invertebrate evolutionary ladder. This elevated position is based on their internal organization, illustrating how deceptive external appearances can be.

The last group to be included in this book are the vertebrates that are found at the shore: fish, birds and mammals. Some fish are accidentally stranded in rockpools, while others are adapted to live on the shore and survive during low water. The birds can be divided into waders, wildfowl and seabirds. The waders and wildfowl are often noted for the variety of adaptions used to exploit the many different food sources available on the shore. As for the seabirds, such as gulls and terns, they often range over great expanses of sea, some scavenging whilst others are specialized divers. Seashore mammals are few: otters and seals.

LIFE CYCLES

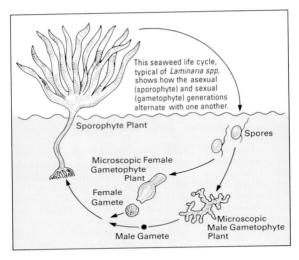

This seaweed life cycle, typical of *Laminaria spp*, shows how the asexual (sporophyte) and sexual (gametophyte) generations alternate with one another.

Sporophyte Plant

Spores

Microscopic Female Gametophyte Plant

Female Gamete

Male Gamete

Microscopic Male Gametophyte Plant

There is one, almost universal characteristic of the reproductive process of all marine species; namely, their dependence on the sea as a 'free-for-all' medium. The sea disperses spores (seeds), gametes (eggs and sperm) and larvae on its currents, acting as an enormous mixing pot and reservoir of potential life, the constituents of which are continually changing. But, the price the sea exacts is the risk of non-delivery; it is in coping with this that the plants and animals of the shore have evolved some fascinating life cycles.

Seaweeds

In its simplest terms, the seaweed life cycle comprises two alternating generations, one asexual one sexual, known as the sporophyte and the gametophyte generations respectively. There are good biological reasons why seaweeds have such apparently complicated life cycles. The intertidal region is a very competitive environment with a high premium being placed on any spare space suitable for colonization. The sporophyte plant is capable of producing truly vast numbers of spores, which it releases into the sea. These spores need only to settle onto a suitable rock to grow into new plants, the male and female gametophytes. The sea is full of such spores; the potential for colonization is forever present. All you need to do is to visit a shore shortly after a storm to see that where boulders have been turned over or old seaweeds torn away you will find new carpets of sporlings, usually of green seaweed.

But the production of spores is an asexual process, there is no mixing of parental genes, no possibility of variation within the species – nothing for evolution to work on. The seaweeds have overcome this problem by alternating the sporophyte generation with a sexual, gametophyte generation. But whilst the chances of any one spore surviving long enough to colonize are incredibly small, the chances of success with sexual reproduction are much

16

less. Not only does a gamete, produced by the gametophyte plant, have to first meet another gamete to bring about fertilization, but the resulting zygote has to survive long enough to colonize. The seaweeds invest a lot of energy in the production of gametes. The prize of variation within the species makes the price worth paying.

There are many modifications to this basic theme. With some seaweeds, for example the sea lettuce, the sporophyte and gametophyte plants are similar in form. In others they are very different, as with the brown seaweeds in the genus *Laminaria*, where the gametophyte plant is microscopic while the sporophyte can be up to four metres in length. In contrast, some brown seaweeds have abandoned the gametophyte generation altogether.

There are various possible reasons for the two generations to develop strikingly different forms: avoidance of competition for space; survival under a range of different seasonal conditions; differing tolerances to grazing pressure; different degrees of competitiveness. In what is a relatively hostile environment, the ability to survive a range of conditions, in some form or other, is clearly an advantage, if not a necessity. Where the two generations have a similar form, it is usually an opportunistic and relatively short-lived species.

Animals

With all the different invertebrates found on the shore, the diversity of life cycles is so great it is perhaps surprising to find a common thread running through many of them – they are adapted to make the most of any space that comes available for colonization. Although some animals (such as sponges, sea anemones and sea squirts) retain the ability to reproduce asexually, most species have abandoned it in favour of sexual reproduction. But the problem of fertilization remains; gametes shed into the sea are rapidly dispersed and diluted before they have a chance to fuse. So the animals have evolved methods to enhance their chances of success.

Synchronization of both the production and release of gametes is controlled by seasonal and environmental factors, such as day-length, temperature and chemical stimuli. For example, when the male edible sea urchin releases its gametes, this provides a chemical cue for females in the immediate vicinity to shed their gametes. Other cues are suspected, such as with the common limpet where violent storms are believed to stimulate mass spawning on a shore.

Once fertilization has taken place, there is usually a free-living, planktonic larval stage. The larvae are dispersed by ocean currents, waiting to colonize new shores, but they also suffer great losses. The animals that you admire on the shore are literally the lucky few.

For any individual plant or animal, its development from a spore or embryo to the moment that it colonizes a shore is only the beginning of its life cycle. Once established, the prime objective is rapid growth, a phase which may last for months or even years. Ultimately, it matures and contributes to the reproduction of the species. When confronted with more than one of the stages in a life cycle coexisting on the shore, it is interesting to reflect on the achievement and the dangers each organism has so far overcome.

POLLUTION AND CONSERVATION

The sea suffers from the old belief that it represents a bottomless well, able to absorb and disperse all that is pumped or dropped into it. As is becoming increasingly evident, this is simply not the case. Pollutants are typically classified according to their source – domestic, agricultural or industrial. They do not have to be harmful in themselves in order to damage the marine environment, excessive amounts of otherwise innocuous substances can do just as much harm as recognized toxins.

Domestic pollution

This comes in two forms, sewage and domestic waste. Large amounts of raw or only partially treated sewage (13.5 million litres a day from the United Kingdom alone) are still pumped out onto beaches and into shallow waters. Even greater amounts (about nine million tonnes of wet sewage sludge a day) are dumped at sea, from where much of it returns to pollute the seashore.

In small amounts, although unpleasant to us, sewage can be beneficial to certain shore-dwelling animals and plants. It contains both inorganic nutrients (fertilizers) and organic material. The organic material is food to many of the filter feeding animals on the

KEY	
1 Antifouling chemicals: used to protect ships	4 Foam: due to detergent pollution
2 Oil slick: often mistaken by seabirds for a shoal of fish	5 Sewage: large amounts are pumped directly into the sea
3 Red tide: the result of algae feeding on sewage	6 Fertilizers: in drainage from farmland

18

shore, such as mussels and worms. Often the largest individuals and most dense populations of these animals are found near sewage outfalls.

The high inorganic nutrient levels in sewage stimulate the growth of planktonic, microscopic algae. The increased growth rate leads to a population explosion, or bloom, which can result in a phenomenon known as a red tide. The red colour is due to the photosynthetic pigment found in one group of algae, the dinoflagellates. Sometimes these dinoflagellates produce toxic chemicals that affect other forms of marine life. When the algae die they sink to the seabed, where they are eaten by bacteria. There are many naturally occurring bacteria that do this job. The sudden abundance of food causes the bacterial population to increase, eventually using up all the available oxygen. The result is an oxygen-free zone, where the sediment is usually black and odorous, and in which very little can live. In addition to these 'harmless' bacteria, others such as *Salmonella* thrive in this environment, presenting a real threat to humans.

Large amounts of detergents are also disposed of in domestic waste. As a result increasing amounts of dirty brown/grey foam can be found trapped in areas of slack water or stranded on the shore. These detergents reduce the aeration of the water, can damage some of the more sensitive species and can increase the effects of certain other pollutants.

Many shores are scarred by the litter of our throwaway society. There was a time when after a storm the tide line might be combed and the occasional wooden fish box, piece of cargo swept overboard or glass bottle would be found. These materials were useful and often collected. Today, the tide line is more likely to be a hydrocarbon junk yard of plastic bottles, polythene wrappers, polypropylene ropes together with aluminium cans and broken glass. Much of this litter is not only unsightly and virtually indestructible but also potentially dangerous.

Agricultural pollution

The runoff from farm land, in the form of dissolved inorganic nutrients from fertilizers, has a similar effect on the marine environment as sewage in that it promotes algal blooms. The point of discharge of this runoff onto the shore is often distinguished by the presence of bright green seaweeds (usually *Enteromorpha spp*). This is due in part to the reduced salinity, but also to the increased nutrient levels.

Industrial pollution

This type of pollution often contains highly toxic chemicals (including radioactive isotopes, heavy metals, biocides and deter-gents, to name but a few) and comes from a vast array of sources including tanneries, the dye industry and petrochemical plants. Once in the sea, these chemicals form a deadly cocktail, which kills marine organisms directly; or more insidiously, the chemicals accumulate within the biological system, appearing in the greatest concentrations in animals at the top of the food chain. Observable effects include the thinning of egg shells and consequent failure of breeding in seabirds. Humans do not escape: in Minemata Bay, Japan, 2,000 people have suffered mental and physical deformities and more than 40 people have died from heavy metal (mercury) pollution produced by a nearby chemical plant, which has been accumulating in shellfish eaten by the local population.

In addition, there are poisons, such as antifouling chemicals, which are intentionally introduced into the sea. These chemicals are extremely toxic, working as a poison at concentrations as small as a few parts of toxin per million parts of water. They are designed to prevent marine animals and plants from growing on boat hulls, fish farms nets and the underwater parts of other permanently moored structures, such as oil and gas platforms.

The chemicals do not stay on the surface they protect, but instead dissolve slowly into the water. This does not present a problem on large ocean-going vessels, which do not remain in one place for any length of time. However, when antifouling chemicals are applied to small pleasure craft, which spend most of their time in marinas, then the concentrations can increase dramatically. The highly toxic TBTs (trybutyl tins) have now been banned for use on small craft in the United Kingdom, but as yet there is not a general EEC ban. New rubber Teflon antifouling paints, which present a surface that is difficult for animals and plants to anchor to, are replacing the toxic variety. They are still slightly toxic, but not as harmful to the environment as TBTs.

Associated with the ever increasing numbers of fish farms in the sea are fears of pollution from nutrient enrichment (due to the accumulation of the fish's wastes products under the cages) and pesticides (used to control fish parasites). One such pesticide is Nuvan which is used to control salmon louse, but unfortunately it is also very toxic to many other crustacea and certain molluscs.

Visually the most devastating and also certainly one of the most globally damaging forms of pollution is oil. An oil spill is doubly damaging; it pollutes when in the open sea and then again when it is washed ashore. An oil slick at sea acts like a magnet to many species of diving bird. Its appearance on the surface of the water deceives the birds into believing it to be a shoal of fish. The birds dive only to be covered with oil. Many of them die from exposure as the oil reduces the insulating properties of their feathers, whilst others are poisoned when they ingest the oil during preening.

If a major oil slick reaches the shore it can cause the virtual total destruction of animal and plant communities. What is most frightening is that what has taken many generations to create can be wiped out almost overnight. The oil coats the large brown seaweeds increasing their weight and drag, causing them to be torn from the rocks by the waves. This jeopardizes the survival of many animals and smaller seaweeds. They may themselves be torn off the rocks as a result of the increased drag, poisoned or simply die without the protection afforded by the larger brown seaweeds.

Despite the scale of devastation, recolonization of the shore starts almost immediately. At first, only a few species are present in very large numbers, but gradually over a number of years the diversity of species increases and with it the complexity of the whole community. A shore in the early stages of recovery is often colonized by the green seaweeds, *Enteromorpha spp*. Next, grazing snails and limpets gradually return, creating areas of bare rock which are colonized by barnacles and brown seaweeds; and so the process continues. A complete recovery may take many years.

Dispersing an oil slick contains its own problems, and the dustbin usually chosen is the sea itself. Dispersants, which are really just powerful detergents, are sprayed onto the oil to break it up into tiny droplets, so that it sinks in the water. Whilst this reduces the threat to diving birds, it simply shifts the problem to the seabed. These dispersants, despite recent improvements, are still many times more toxic than the oil itself. Their application to an already weakened biological system may often do more harm than good.

Dispersing the oil reduces the risk of it coming ashore, but unless the threatened shoreline is particularly vulnerable, allowing the oil to beach is often the lesser of two evils. Leaving oil to weather naturally (there are a number of bacteria that feed on oil, and the more toxic fractions quickly evaporate off) is often the more ecologically desirable choice.

Heat pollution can also damage the seashore environment; for example, when seawater is used as the coolant in a power station. A small increase in the temperature of the discharged water, when compared to the surrounding seawater, will result in nothing more than a slight acceleration in the growth rates of most organisms around the outfall. However, a critical point is soon reached at which some species thrive whilst others are killed. The common mussel thrives in warm waters whilst its main predators, the dog whelk and the common starfish, are killed.

Conservation

It is essential that we continue to utilize the resources of the sea, whilst keeping in mind the impact of our various activities. Conservation measures on an international scale are necessary and organizations such as the World Wild Fund for Nature, the Marine Conservation Society and the National Trust work towards that end. These and other conservation bodies need our support if they are to be successful (see below).

At a personal level there are a number of steps that you can take. By becoming aware of how the products you use in your home are disposed of, and selecting environmentally sound alternatives where they exist, you can reduce the pollution caused on your behalf. Active support for local campaigns aimed at improving the marine environment is also useful. When visiting the shore, take care to ensure that damage does not result from your enjoyment by replacing animals and plants once you have examined them.

Friends of the Earth, 26-28 Underwood Street, London N1 7JQ
Greenpeace, 30-31 Islington Green, London N1 8XE
Marine Conservation Society, 9 Gloucester Road, Ross-on-Wye, Hereford HR9 5BU
National Trust, Enterprise Neptune, 36 Queen Anne's Gate, London SW1H 9AS
Nature Conservancy Council, 19 Belgrave Square, London SW1X 8PY
World Wide Fund for Nature, Panda House, Weyside Park, Godalming, Surrey GU7 1XR

CLIFFS, SPLASH-ZONE AND SALTMARSHES

In the minds of many people, a visit to the seashore starts when the tide line is crossed and the first pieces of live seaweed are encountered, or an expanse of sand or mud is reached. However, the influence of the sea extends well above the high-water mark to adjacent cliffs, the splash-zone and saltmarshes, which are covered only by exceptionally high tides.

Cliffs

If the tide is in, there can be no more relaxing or pleasurable way of passing the time than watching the acrobatics of the seabirds as they ride the upwelling air currents or thermals, which are common near cliffs. The inaccessibility of many high cliffs makes them

KEY	
1 *Sterna albifrons* Little Tern	5 *Ramalina siliquosa* Sea Ivory
2 *Silene maritima* Sea Campion	6 *Anaptychia fusca*
3 *Armeria maritima* Thrift	7 *Fratercula arctica* Puffin
4 *Xanthoria parietina*	8 *Verrucaria maura*

excellent nesting sites, protected from predators. With a pair of binoculars you can explore them from a safe distance. Moreover, cliffs are often rich in flowering plants.

Many of the gull-like birds found nesting in potentially accessible places are fulmars. These birds are often seen skimming the crests of the waves and are undisputed masters of the thermals. It is not a good idea, however, to try to get a really close look at a nesting fulmar, for apart from the disturbance caused to the birds, fulmars have a very nasty habit of 'spitting' an oily and smelly discouragement over considerable distances with impressive accuracy.

Members of the auk family are more noticeable for their sheer numbers than their soaring flight, although the apparently kamikaze flight of puffins as they head for what is almost a vertical cliff, only to stall and disappear into their nest burrows, is something to be wondered at. The crowded 'nest' sites (no nest is actually built) of guillemots are fascinating to watch, both for the birds' skill in incubating an egg laid on an almost non-existent rock ledge and for the ability of individuals to return time after time to the same spot on the cliff face.

The season for nesting birds is short, lasting from late April through to early August. During this period the cliffs are alive with the noise of continually squabbling birds and their nonstop flights in search of food. Often great rafts of resting birds can be seen sitting on the water below the cliffs.

Where the cliffs are not precipitous and inaccessible there is the opportunity to study the flowering plants at close quarters. In late spring and summer the clifftop can offer an exhilarating combination of panoramic seascapes and a riot of colour from flowering thrift, sea campion, rock samphire and many other species. It is possible to identify many of the plants even when not in flower. The environment in which they are found is an arduous one. Cliffs are subject to many extremes; erosion, drought and flooding. Naturally, the species found here are adapted to the conditions. The surface area of their leaves is either reduced or covered in a waxy coat to minimize water loss and salt spray damage. The root system is large and woody to provide as secure an anchor as possible against erosion and wind. However attractive the flowers, resist the temptation to dig up a clump of thrift or any other plant for the garden; not only is it unlikely to survive, it is also against the law.

Splash-zone

This is where terrestrial vegetation meets the marine influence. The splash-zone, dominated by lichens, is best developed and most extensive on more exposed rocky shores. It is an inhospitable habitat and as a result has relatively few species, but it is definitely worth more than just a passing glance.

It is here that you will find many lichens, which are a fascinating and beautiful group of plants. They have a number of different growth forms, ranging from the encrusting black lichen *Verrucaria maura*, through the low-growing, crumb-like *Caloplaca marina* and the leaf-like *Xanthoria parietina*, to the delicately robust, erect *Ramalina siliquosa*.

There are also a few flowering plants, such as thrift and sea campion, which tend to occur in the more sheltered crevices, where a little soil can collect. It is worth searching amongst the lichens and the flowering plants and in cracks and crevices for the animals that may be present. These include the small black snail *Littorina neritoides*, the sea slater and a few marine insects. *L. neritoides* is virtually terrestrial, but it is still dependant on the sea for its reproduction. The sea slater is a scavenger, normally venturing out only at night and hiding by day in damp cracks and crevices, which it shares with the bristle-tail.

The size of the splash-zone gives the first clues as to what may be found further down the shore. A wide zone, with abundant lichen cover, is indicative of an exposed shore and most likely to have a limpet/mussel/barnacle zone in the mid-tide region and the brown seaweeds, thongweed and dabberlocks, below. A narrow or non-existent splash-zone is indicative of shelter and generally associated with egg wrack covered shores.

Saltmarshes

Characteristic of more sheltered areas, saltmarshes are found mainly in estuaries, on the landward-facing shores of islands and in sheltered bays and sea lochs. These places are ideal for the deposition of the fine, organically rich sediments on which this type of marsh develops.

The dominant plants are salt-loving flowering plants, which colonize and stabilize the alluvial muds. There is a definite pattern of zonation of plant species, determined by the frequency of tidal flooding. The uppermost regions of a saltmarsh are almost terrestrial and are only flooded by the very highest equinoctial spring tide. The lowermost regions are covered by the tide on all but the smallest neap tides.

Erosion is one of the greatest threats facing saltmarshes. The overall health of a marsh can be assessed by the extent of the lowest zone of colonization, known as the pioneer zone. Initially the mud is colonized by eelgrass and green seaweeds, such as *Enteromorpha* spp. These plants increase the rate of deposition and thus raise the level of the mud. Once the mud reaches a height at which it is exposed successively over a number of neap tide cycles, the true saltmarsh pioneer species can invade and become established, stabilizing the marsh.

Glasswort is the main pioneer species and can withstand being immersed for long periods. It is ideally suited to the habitat with short, upright stems and branches with curved edges, which reduce resistance to water currents. In young saltmarshes, glasswort forms dense forests which are extremely effective at trapping more sediment. It shares the pioneer zone with a number of seaweeds, notably channel wrack and spiral wrack. Whereas glasswort has a shallow root system for anchorage, the seaweeds must be attached to a stone or boulder to survive. As the mud rises and becomes more stable, additional plant species are able to colonize, such as sea aster and seablite.

The process of deposition continues slowly, and with the mud being exposed for ever longer periods of time seeds of other species

are able to germinate and grow. The typical flora of the mid and upper saltmarsh zones includes such species as sea lavender, thrift, sea purslane and scurvy grass. It is best to visit a saltmarsh in late summer or early autumn as many of the plants are annuals, or if perennial die back each winter.

Large saltmarshes are often crisscrossed by deep drainage gullies. These in turn are blocked by further deposition to form isolated ponds where marine, mud-dwelling animals such as the small snail *Hydrobia ulvae* can survive before the glasswort establishes itself, restarting the sequence of colonization.

Plants are not all that saltmarshes have to offer, they are also excellent areas for bird watching. The extensive mudflats which often occur below saltmarshes provide abundant feeding for waders and wildfowl. The saltmarsh itself is both an ideal roost site for various waders at high tide and a source of food for some ducks. Roosting species include curlew, oystercatcher, redshank and godwit. The low-growing vegetation provides some protection for these birds, without obstructing their view of the surrounding land, which is essential to guard against predators. Duck species, such as wigeon, feed on the numerous seeds that are produced by the plants. Although the saltmarsh vegetation provides good cover for breeding birds, such areas are seldom used as nesting sites due to the effects of flooding.

KEY	
1 *Salicornia europea* Glasswort	4 *Suaeda maritima* Annual Seablite
2 *Zostera marina* Eelgrass	5 *Haemotopus ostralegus* Oystercatcher
3 *Numenius arquata* Curlew	6 *Ameria maritima* Thrift

ROCKY SHORES

*I*t is a good idea to start any visit by finding a convenient vantage point from which to get an overall impression of the shore. A quick scan enables even a relatively inexperienced eye to decide which are the dominant zone-forming seaweeds and animals. It is also useful in locating the different sub-habitats, such as rockpools, boulder areas and overhangs. It is often in these places that the more delicate and less common species are to be found.

A rocky shore at low water is, for a large part, an ecosystem in limbo. Whilst the seaweeds and many of the animals are obviously adapted to survive these conditions, they are really designed to function when the tide is in. Part of the challenge, therefore, is to try to imagine the shore life with the tide in. A glimpse of this is, of course, available in rockpools (see pages 30–33).

Probably the greatest risk to all intertidal organisms is that of drying out when the tide retreats. Living or hiding in cracks and under boulders is the only way many of the more delicate species survive. Other animals, such as limpets, have a hard shell that provides some protection, although those exposed for long periods are still at risk. They reduce this danger by always returning to the same spot after each high tide, to ensure a tight seal between their shell and the surface. On hard rocks the margin of the shell grows to fit the contours of the surface, while on softer rocks, particularly shales and sandstones, the edge of the shell wears an impression in the rock, known as the home scar. Vacant home scars of long since dead limpets can often be seen on the rocks.

KEY	
1 *Patella vulgata* Common Limpet	4 *Cancer pagurus* Edible Crab
2 *Fucus serratus* Toothed Wrack	5 *Mytilus edulis* Common Mussel
3 *Fucus vesiculosus* Bladder Wrack	6 **Semibalanus balanoides** **Acorn Barnacle**

Space to live is at a premium on rocky shores. Whenever it becomes available, green seaweeds, such as *Enteromorpha spp*, are often the first to colonize, producing bright green patches on the shore. These seaweeds do not persist for long, they are soon eaten by barnacles or limpets, or replaced by larger brown seaweeds.

Bare areas of rock can appear and persist in some circumstances; for example, where a few large brown seaweeds have been interspersed with barnacles, which have subsequently died. These seaweeds, as a result of wave action, prevent new colonization of the rock within an area described by the arc of their sweep. The result is that individual plants are surrounded by strikingly bare, roughly circular areas of rock.

There are many signs on the shore of otherwise unseen activities of the intertidal animals. The wanderings of snails are recorded on the rock surfaces. They move by gliding over a thin film of mucous, which they secrete themselves and which acts like a fly paper, trapping any fine sediment suspended in the water. This results in the trail of the snail being highlighted as a light track of silt particles against the darker rock. Try mapping these tracks and measure the distance travelled by individuals in the space of a high tide. By following the trail carefully the animal responsible for making it can often be identified.

Another type of trail is left by the common limpet. The limpet feeds by scraping off the film of microscopic algae, bacteria and filamentous seaweeds from the rock surface. Where this film contrasts with the underlying rock, the limpet leaves a perfect record of the small zigzag sweeps of its mouth as it moves over the surface. Studying these scrape marks can not only confirm whether an individual limpet has returned to where it started from at the beginning of the high tide, but also the angle between the zig and the zag indicates how fast it was moving. The greater the angle the faster the limpet.

Zonation on rocky shores

On sheltered shores the various zones are defined by the large brown seaweeds. Channel wrack occurs at the top of the shore, with a band of spiral wrack below it and lichens in the splash-zone above. The widest zone is in the middle of the shore and is composed of bladder wrack and egg wrack, with the latter proportionally more abundant on more sheltered shores. Towards the low-water mark there is a band of toothed wrack, which eventually gives way to the kelp zone, with sugar kelp and tangle.

With increasing exposure the dominance of the brown seaweeds is reduced. The splash-zone is broader and dominated by various lichens, while the luxuriant growths of bladder wrack and egg wrack, found on sheltered shores, are either much reduced or sometimes replaced by barnacles and/or mussels with limpets. Towards the lower end of the shore the band of toothed wrack is interspersed with additional species, such as thongweed and carrageen. The sugar kelp is also less abundant.

On the most exposed rocky shores the seaweeds are much reduced. The splash-zone is often very large and below it there is sometimes a band of laver. This red seaweed is seasonal and so not always present. The channel and spiral wracks are very much reduced and there is usually a broad barnacle/small mussel zone. The lower regions of the shore are still seaweed dominated, but this time with a mosaic of species including pepper dulse, carrageen, Irish moss, thongweed and dabberlocks.

In a textbook world, one might expect a rocky shore profile to consist of a series of perfectly delineated zones of animals and plants – the reality is more complex. It is often interesting and instructive to try to identify the causes of any variations from the standard pattern. Sometimes these anomalies are easily explained by an undulating shore profile; for example, when comparing two adjacent sections of a shore, you may find that the spiral wrack zone in one is below the bladder/egg wrack zone in the other, simply because the two sections have different profiles. On other occasions the cause and effect may be rather more distanced from one another. The effect of an offshore reef, creating an area of shelter on an otherwise exposed shore dominated by limpets and barnacles, can result in an area dominated by brown seaweeds. In general, the diversity of both animal and seaweed species increases both down the shore and with increasing shelter.

The major zone-forming species are often overpoweringly visible. They cannot be ignored, but neither should they divert attention from the less obvious species. On areas of bedrock it is a simple matter of getting down on hands and knees and examining the ground closely. The diversity of species to be found in cracks, crevices and around the holdfasts of seaweeds is impressive.

Boulder shores

On most rocky shores there are likely to be both areas of bedrock and of boulders and pebbles. The boulder areas often have more species because of the greater diversity of habitats and the fact that they are often more sheltered. All the surfaces of a boulder are available for colonization, along with the gravel or sand sediment in which they are embedded. The upper surfaces are often dominated by large, luxuriant growths of brown seaweeds, in particular egg wrack. The diversity of animal life under such thick seaweed is usually limited; limpets and winkles predominate. The underside is more or less the exclusive domain of animals as there is usually not enough light for plants.

If the boulders are embedded in mud, the mud acts as a very effective seal around the base, preventing oxygen-rich seawater from penetrating the sediment. The resulting oxygen-free conditions are unsuitable for most animals. On the other hand, if the boulders are embedded in sand or gravel then a plentiful supply of oxygenated water is available. Where the boulders are particularly stable, and the rate of water percolation is good, the under-boulder fauna can be quite spectacular.

Some of the more unusual, interesting and delicate animals are only found on the undersides of boulders or in the sediment beneath. Many of these animals are sedentary and could only tolerate brief exposure to the drying conditions experienced on the upper surfaces. If they are left exposed for too long many of them will quickly perish. Similarly, many of the animals and more

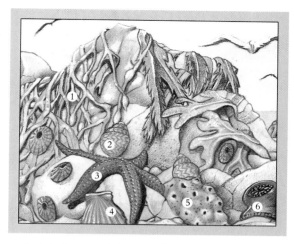

particularly the seaweeds found on the upper surfaces would find the conditions on the underside impossible. So when you move a boulder in order to have a look underneath, it is particularly important to return it to its original position afterwards. Take care when turning boulders over, beware the sharp edges of the tubes of the aptly named keel worms and the shells of the barnacle *Verruca stroemia*. These can easily inflict nasty lacerations to the fingers, particularly when the boulder resists being turned.

Invertebrates are abundant on boulder shores. There is often a bewildering array of encrusting animals including many sponges, sea anemones, various types of tube-dwelling worms, molluscs (including the saddle oyster and scallops, which are often overgrown with sponges) and a variety of sea squirts. Mobile species are equally well represented with scale worms and crabs, including shore crabs and porcelain crabs. There are also many different echinoderms to be found with species of starfish, brittle star and sea cucumber all being present, together with various molluscs including chitons and numerous snails. The sediment is a rich source of worms, both segmented and unsegmented, such as the amazingly long bootlace worm and segmented bristle worms. A list such as this is merely the tip of the iceberg and serves only to illustrate the spectacular diversity of life that is waiting to be found on boulder shores.

KEY	
1 *Ascophyllum nodosum* Egg Wrack	4 *Pecten maximus* Great Scallop
2 *Monodonta lineata* Thick Topshell	5 *Myxilla incrustans*
3 *Asterias rubens* Common Starfish	6 *Actinia equina* Beadlet Anemone

ROCKPOOLS, CREVICES AND OVERHANGS

Rockpools are natural aquaria, occurring in all shapes and sizes on most rocky shores. They vary from large, shallow pools, often with pebble or sand bottoms, to small, very deep holes. Most people only see the plants and animals of a rocky shore when the tide is out. A few, those who have put on wet suits and Scuba gear, are privileged to have observed the inhabitants of the shore in their preferred state, namely, with the sea covering them. For most people, rockpools are the only opportunity to observe marine life in a truly aquatic environment in the wild.

Crevices and overhangs provide intertidal species with shelter from the stresses of the environment. Cracks, fissures and even small depressions in the rock can make a considerable difference to the chances of survival of common shore species. In deeper crevices whole new, specialized, zoned communities can be found.

KEY		
1 *Patella vulgata* Common Limpet	5 *Eupagurus bernhardus* Common Hermit Crab	
2 *Littorina littorea* Edible Periwinkle	6 *Membranoptera alata*	
3 *Ulva lactuca* Sea Lettuce	7 *Asterias rubens* Common Starfish	
4 *Blennius ocellaris* Butterfly Blenny	8 *Mytilus edulis* Common Mussel	

Rockpools

The plants and animals of the upper shore are adapted to the stresses of being out of water for long periods (maybe up to ten hours at a time). They have to tolerate large changes in salinity, temperature and the risk of drying out. In contrast, in a rockpool on the upper shore there is no problem with drying out and the changes in both salinity and temperature are much less. In effect, living in a rockpool is much more like living near or below the low-water mark. As a result, animals and plants that are usually to be found on the lower shore occur in rockpools. A good example of this are the limpets, *Patella vulgata* (common limpet) and *Patella ulyssponensis*. The common limpet is the dominant species over much of the shore, particularly in northern latitudes, while *P. ulyssponensis* is restricted to the bottom of the shore. In pools, however, *P. ulyssponensis* often predominates regardless of the shore level of the pool.

As the animals and plants found in a rockpool high up the shore could not survive so far from the low-water mark if they were removed from the pool, you should always carefully replace anything that you take from a pool for closer examination.

When approaching a rockpool, it is a good idea to move slowly and with care. Vibrations set up by movement over the shore are readily transmitted through the rock and the water frightening the animals. If at all possible the pool should be approached into the sun to avoid casting shadows over the water. If on first inspection the rockpool appears lifeless, then just stop and quietly observe for a few minutes. The chances are that the inhabitants will resume their activities and your patience will be rewarded.

A typical rockpool community is a collection of seaweeds and animals. The seaweeds are attached to the sides and bottom and members of all three major groups (red, brown and green) may be represented. As for animals, look out for sponges, sea anemones, molluscs, starfish, sea squirts and small fish.

At first sight, many rockpools have what appears to be shell debris on the bottom, but in fact many of these shells are occupied. Those still being used by their original owner make slow progress in search of food, gliding over a stream of mucous on a single foot. If you pick up such a snail it will rapidly withdraw its foot for protection. Amongst the many shells there may be some that can be observed scuttling around at a much faster rate. These have new tenants, hermit crabs, which use empty shells for protection as they scavenge for food.

As snails grow they are able to add to their shell in order for it to accommodate their increase in size. Hermit crabs, on the other hand, only have squatters' rights and as the crab grows it gradually becomes too large for its shell. To continue to grow it must move home. There is never any shortage of vacant dwellings, but the correct shell must be carefully selected. It is important that the shell it selects is not only large enough to accommodate it at the present, but also provides room for it to grow. Crabs 'thinking' about moving can be seen on the bottom of the rockpool, picking up shells with their large pincers and examining them for size. If you are lucky you might see a crab transferring from its old shell to a new one. This is when the soft-bodied crab is at its most vulnerable. With the new shell strategically positioned the crab quickly vacates the old shell and takes up residence in the new.

Many of the simpler animals, such as sponges and sea anemones, favour rockpools. A number of different species of both these groups of animals occur on the shore. Careful identification is needed, especially as there are often many different colour variations within a single species, which can confuse the unwary. If exposed at low tide, sea anemones appear as blobs of jelly-like material, as they are forced to withdraw their tentacles and await the return of the tide. In rockpools, however, they are able to remain actively feeding regardless of the state of the tide, by extending their crown of tentacles into the water. When a particle of food touches a tentacle it triggers a mechanism which fires a poisoned dart attached by a fine thread. The speared prey is reeled in and passed down the tentacle to the mouth which is situated in the centre of the crown.

The sea anemone's tentacles present a tasty morsel when waving around in the water, to various species of fish. The anemone has evolved a defence reaction, rapidly retracting its tentacles in response to attack. This is easily demonstrated by touching a single extended tentacle; the anemone will retract it, and if you persist in touching adjacent tentacles the response will spread rapidly until the whole crown of tentacles is withdrawn.

Rockpools also give you the opportunity to observe seaweeds in their natural growth form, supported by water. Red seaweeds in particular display much truer coloration in pools since they are not bleached by exposure to the sun.

Some patience and an enquiring mind are required to get the most from a rockpool, and you should always be ready for the unexpected. Very often fish or animals which normally live on the seabed, such as sea urchins, can be stranded in a rockpool by the receding tide.

Crevices and overhangs

On bedrock shores, small crevices and cracks in the rock surface often stand out because of the animals and plants which congregate there. Particularly on exposed shores, the protection from wave action provided by a crack attracts winkles and dog whelks. Not only do they gain some protection from dislodgement, but the atmosphere remains considerably more humid than elsewhere, thus reducing the threat of drying out during periods of low tide.

Broad, shallow runnels in the rock surface are also highlighted by the communities they support. These areas are continually damp, enabling various quite delicate seaweeds, such as pepper dulse, *Corallina officinalis* and *Leathesia difformis*, to grow in dense carpets. These seaweeds in turn have associated populations of animals, including small snails and crustacea. The edges of the pepper dulse frond are a favourite place for young limpets, which thrive in the damp environment under the seaweed, protected from predators and the bulldozing activity of larger limpets.

Deep crevices often support quite complex communities. With increasing depth there is less light, insulation from the extremes of

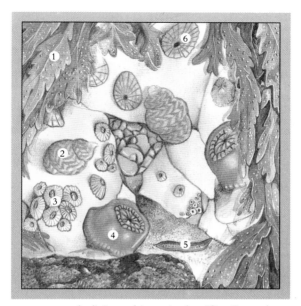

temperature and salinity, and increasing humidity. Seaweeds are confined to the upper margins of the crevice, where there is sufficient light for photosynthesis. They can act as a barrier, trapping moisture when the tide is out, keeping the crevice environment damp enough for animals that would otherwise only be found near the bottom the of the shore or in rockpools. The walls of a crevice are colonized by the more delicate, filter-feeding animals, such as sponges, sea anemones, sea mats, sea firs and sea squirts; there are also various grazing molluscs. At the bottom, there is often an accumulation of rich organic sediment, which provides a home for worms, molluscs and sea cucumbers.

Under rock overhangs you are likely to find a dense and diverse fauna. The upper surface of the rock is often covered in seaweeds, which hang down concealing the overhang at low water, creating a hidden environment of high humidity, low light and shelter from both wind and waves. These conditions are ideal for many of the more delicate animals. When the seaweeds are pushed aside, you are likely to find sponges (in particular the purse sponge), sea anemones, sea squirts and a variety of molluscs (including various species of sea slug).

KEY	
1 *Fucus serratus* Toothed Wrack	4 *Actinia equina* Beadlet Anemone
2 *Gibbula cineraria* Grey Topshell	5 *Capitella capitata*
3 *Semibalanus balanoides* Acorn Barnacle	6 *Diodora graeca* Keyhole Limpet

SANDY AND MUDDY SHORES

Compared to rocky shores, sandy and muddy shores appear less promising for the naturalist; however, all you need to do is to approach them in a different way. The uniformity of a sandy or muddy shore does mean that there is not the same diversity of habitats as on a rocky shore, and so fewer species.

The nature of a particular shore depends on how exposed it is to the waves. This determines the size of the particles that make up the shore, which can vary from coarse sand on exposed shores, to fine muds and silt on the most sheltered. You can gain many clues as to the type of sediment from looking at the surface (see pages 12–13); each type supporting its own characteristic community of animals and plants, and like rocky shores there is a definite zonation of species. The standard zonation pattern for both exposed (sandy) and sheltered (muddy) shores is illustrated on page 10.

With increasing shelter the organic content of the sediment rises. To begin with this results in better feeding for many of the burrowing animals. But the organic matter also provides food for bacteria which live in their billions in the sediment. As the bacteria multiply, they ultimately consume all the available oxygen, producing a completely oxygen-free layer in which only a few highly specialized animals can survive. You can identify this layer by its colour; it is grey or black and can occur at varying depths below the surface. In the most extreme cases it also has a characteristic bad egg smell.

In general, it is necessary to dig to discover the inhabitants of a sandy or muddy shore. Usually you need to sample only the top ten centimetres, as most of the animals live near the surface. But there are a few species for which it may be necessary to dig down 50 centimetres. In any sample there is going to be a large amount of sand or mud relative to the animals present. This needs to be reduced in order that the animals may be more easily examined. Ideally, the sample should be sieved through a garden riddle with a mesh size of between one and two millimetres. This allows all but the largest particles to pass through, while retaining all but the smallest animals. Then, to view the animals against a contrasting background, tip the remaining material into a white tray or bucket containing a little seawater.

An alternative but less satisfactory method is to put the sample in a bucket of seawater, stir vigorously, and then to quickly decant the water containing the smaller animals into a second bucket. You can then retrieve the larger worms and the molluscs by hand from the first bucket.

In most samples there will be many small worms and crustacea. These require expert knowledge and a good microscope before they can be identified, but it is still worth taking some time to observe their numbers and diversity.

The larger seaweeds (e.g. brown seaweeds) are, for the most part, absent from sandy and muddy shores due to the lack of suitable boulders or rocks to which to attach themselves. However, some of the smaller, more delicate species can survive attached to pebbles, particularly on more sheltered shores.

Sandy shores

A little sharp-eyed detective work will provide the first clues to the animals hiding beneath the surface. The lugworm advertises its presence by a coiled mound of sand on the surface, known as a cast, the size of which is roughly proportional to the size of the worm. It is found from the middle shore downwards. The lugworm lives in a u-shaped tube, with the tail end of the tube beneath the cast. It draws sand particles into the head end of its tube when the tide is in. After digesting the bacteria and other organic matter sticking to the grains of sand, the worm excretes the sand as the tide recedes, producing the cast at the tail end of the tube. It is able to live in the black, oxygen-free layer (see above) because it draws oxygenated water from above.

Many of the bivalve molluscs living in the sand are relatively easy to collect since they do not normally burrow very deeply. Different species have developed a variety of feeding methods. The edible cockle feeds off particles of food suspended in the water and can occur in very large numbers, buried only a few centimetres below the surface. When the tide is in, the cockle extends two short tubes of equal length, known as siphons, from between its gaping

KEY	
1 *Angulus tenuis* Thin Tellin	5 *Fabulina fabula*
2 *Echinocardium cordatum* Sea Potato	6 *Ensis siliqua* Pod Razor Shell
3 *Cerastoderma edule* Common Cockle	7 *Arenicola marina* Lugworm
4 *Mya arenaria* Sand Gaper	8 *Haemotopus ostralegus* Oystercatcher

shells, through the surface of the sand. The inhalant siphon has a frilly margin which filters out large particles. The oxygen-rich, plankton-laden water is then passed over the gills, which filter out suitable food particles and absorb oxygen before the water is passed out through the exhalant siphon.

The thin tellin is a common inhabitant of the lower regions of sandy shores. It feeds on particles of food that have been deposited on the surface of the sand using two, long, separate siphons. The inhalant siphon hoovers the surface of the sand in an arc, picking up particles of organic matter. The shorter exhalant siphon ejects the waste material into the water. The hoovering action creates a shallow depression in the sand.

Unlike the cockle and the tellin, the razor shell, which is to be found in the lowest regions of the shore, can burrow both fast and deep. It is extremely sensitive to vibration, detecting even the most stealthy approach. Fast spadework is necessary to catch them as their streamlined shape disappears deep into the sand; and when you do, beware, it is not called a razor shell without reason – not only do they resemble an old cutthroat razor in shape, but also the edges of the shell are sharp. In the lower regions of the shore you will also find specialized worms such as the sandmason worm, which forms dense forests of sand tubes, sticking up above the surface of the sand.

To avoid random digging, look at the water's edge on a receding tide. Many of the animals continue feeding right up until they are exposed to the air; telltale plumes of sandy water or protruding siphons can often be seen. Stealth is essential because, like the razor shell, the animals are sensitive to vibration. You will often find birds patrolling these same water margins, looking for a juicy meal in the form of an extended siphon or tentacle, which can easily be nipped-off.

A few species of echinoderm have adapted to living on sandy shores. One of these is the sea potato; the empty, bald tests (hard, outer casing) of which are often found washed up on the shore. This is all that is usually seen of these animals. When alive the test is covered in spines and the animal burrows through the sand at a depth of 20 centimetres right at the bottom of the shore, using specialized tube feet to collect food particles.

Muddy shores

You will often find muddy shore species in very large numbers, feeding on the highly organic sediment. They have to be able to cope with their feeding mechanisms being clogged by the very fine mud particles. Many of them are burrowers, but not as mobile as their sandy shore counterparts.

The most common deposit feeding bivalves on the muddy shore are the baltic tellin and the peppery furrow shell. The baltic tellin is the muddy shore equivalent of the thin tellin found on sandy shores. It is a more rotund animal, with less need to burrow through the sediment in search of good feeding.

The gaper is a suspension feeder and one of the largest burrowing bivalves. It lives deep in the mud, but compared to the razor shell it is easy to catch as it is not capable of rapid movement.

The edible mussel is a surface dwelling, suspension feeding bivalve. It forms large mussel beds, which are a common feature of many muddy shores. Once established, these colonies are self-perpetuating; the older mussels and empty shells providing surfaces on which new mussels settle and grow. This results in a hard, stable surface on an otherwise unstable shore, which often supports an increased diversity of species. Seaweeds can grow together with some gastropods typical of rocky shores, such as the edible periwinkle, the flat periwinkle and the dogwhelk. There is one gastropod, the layer spire shell, which is adapted to living on the mud surface.

Many other intertidal species also occur in high-density colonies, and it is this that makes muddy shores such attractive feeding sites for waders and wildfowl.

Tide line

This feature is characteristic of all types of shore, but most easily examined on the flat expanse of a sandy or muddy shore. It is an intriguing clutter of natural and manmade flotsam and jetsam that awakens the beachcomber in most of us. The tide line is a source of otherwise inaccessible deeper water species, and the remains of more common animals and plants from intertidal regions. In addition, you will often find that much of it has originated from storm damage on another part of the coast. Sadly, more and more commonly, this natural line of debris is scarred by man's indestructible rubbish.

The bulk of the tide line is generally a tangled mass of seaweeds including brown seaweeds, kelps and dead mens' ropes. Many stranded animals are there as a result of their intimate association with the seaweeds. These unfortunate animals must perish along with the plants to which they are attached; one of the most attractive of these is the blue-rayed limpet, attached to kelp. Other animals include corpses of fish, birds, many different invertebrates and sometimes the carcass of a seal, dolphin or whale.

Occasionally the tide line shows evidence of some catastrophic mortality; for example, when a strong onshore wind drives shoals of sand-eels and jellyfish ashore, leaving them to die. Such natural disasters provide rich pickings for scavengers, such as gulls, which take anything from dead snails to abandoned half-eaten sandwiches.

One animal that has become specially adapted to the tide line environment is the sandhopper. This crustacean feeds on the rotting seaweeds and is often present in large numbers. It has adapted to the almost terrestrial habitat and can tolerate large fluctuations in both salinity and temperature. Also found amongst the seaweeds is the aptly named turnstone, a small bird which methodically searches for small invertebrates, including the sandhopper.

The tide line can serve as a guide to the size of the tide, particularly during the period following a spring tide. At this time you will often find a number of parallel tide lines; the uppermost line represents the height of the spring tide, the lower lines those of smaller tides.

ESTUARIES

Fluctuating salinity is the key factor affecting the flora and fauna of an estuary. The area over which the freshwater and seawater mix varies depending on the size and flow of the river and the state of the tide. With a small river the freshwater may be confined to the river mouth; with a large river, in flood, the salinity may be reduced for a considerable distance away from the river, along the coast. Whatever the size of the river, the range of the seawater effect (up the river on the flood tide) and the freshwater effect (down the river on an ebb tide) depends on whether it is a spring or neap tide. Typically, the seawater will penetrate much further upstream (and the freshwater downstream and out to sea) on a spring tide.

An estuary provides an excellent opportunity to observe how environmental factors control the distribution of different species. It represents an easily defined, relatively uniform (albeit fluctuating), environmental gradient (from freshwater, through a 50:50 mix, to seawater) against which the relative distribution of species can be mapped.

One group of closely related species that neatly illustrates this are the three shrimps belonging to the genus *Gammarus*. The freshwater species is abundant in the non-estuarine regions of the river, but extends only a very short distance into the area of

KEY	
1 *Haemotopus ostralegus* Oystercatcher	5 *Fucus ceranoides* Horned Wrack
2 *Halimione portulacoides* Sea Purslane	6 *Salicornia europea* Glasswort
3 *Anguilla anguilla* Common Eel	7 *Buccinum undatum* Common Whelk
4 *Enteromorpha spp*	8 *Mytilus edulis* Common Mussel

saltwater influence. The brackish water species dominates the area in which major salinity changes occur, whilst the common shore species, which is abundant along the open coast, only extends into the estuary a very short distance.

The tolerance of different seaweeds to both the reduced salinity and the increased murkiness of estuarine waters is, to a certain extent, a reflection of their zonation on an open rocky shore. The red seaweeds are the least tolerant of reduced salinity and rarely extend into an estuary. The brown seaweeds penetrate further, but quickly decline in diversity, with toothed wrack the first to disappear. Bladder wrack can be found up to the middle regions of the estuary, while horned wrack, which is a truly brackish water species, is characteristic of estuaries. The green seaweeds, such as *Enteromorpha spp*, extend the furthest up an estuary.

Most of the estuarine animals are mobile to some degree, but unable to move far or fast enough to escape the rapid changes in salinity with each tide. Many common species can be found considerable distances up river estuaries including the edible winkle, the common limpet and the shore crab. Although they can tolerate reduced salinity, it is often at the expense of stunted growth, with many of them displaying dwarfism. Their capacity to reproduce is also affected.

Typically estuaries are fringed by reed beds with saltmarshes. Many estuaries have very extensive areas of sandbanks and mudflats, which are formed from the sediment carried downstream in the river and deposited where the river slows, as the fresh and saltwater mix. These banks are continually being deposited (and eroded) and their position slowly changed due to subtle variations in the routes of the deep water channels.

The mudflats support large numbers of a few specialized animal species, which feed on the organic matter deposited with the alluvial mud. The main species are the lugworm, the ragworm, the layer spire snail, the baltic tellin, peppery furrow shell and the amphipod crustacean *Corophium volutator*. The edible mussel is often found in large mussel beds, extending over expanses of the intertidal flats of an estuary.

All these animals are the favoured food of large numbers of waders and wildfowl. It is this massive food resource which transforms what would otherwise be a bleak environment into a place teeming with life, with literally thousands of overwintering birds feed there during the winter months. Each estuary attracts its own unique collection of bird species, which are adapted to feed at slightly different depths in the mud or on different prey species.

An estuary is not the only place where changes in the salinity play a major role in the ecosystem. Many intertidal organisms, on all types of shore, have to cope with variations in salinity. For example, when the tide is out, rain can dilute the seawater in a rockpool or gully, dramatically reducing the salinity; alternatively, strong winds and sun increase evaporation which raises the salt level. However, these changes occur at irregular intervals and relatively infrequently and can be tolerated by most intertidal animals and plants. The estuarine environment, in contrast, experiences regular fluctuations.

THE SHORE
AT HIGH TIDE

Snorkelling in the shallow waters that cover the shore at high tide reveals the intertidal animals and plants in their preferred, aquatic environment. When exposed by a receding tide, much of the shore life is suspended, waiting for the returning tide to reactivate the ecosystem. The inhabitants of the shore are, of course, adapted to survive these periods of exposure to air, but for the animals to feed and the plants to photosynthesize they need to be immersed. So, when visiting the seashore, take every opportunity to walk along the waters' edge or to look down into deeper water. Only the briefest of descriptions of the type of transformations that take place under water is possible here.

The sea is seldom calm, and the rise and fall of the tide is often associated with at least some wave action. The seaweeds are tugged and pushed in all directions by the breaking waves and undertows,

KEY	
1 *Laminaria hyperborea* Cuvie	5 *Asterias rubens* Common Starfish
2 *Palmaria palmata* Dulse, Sheep's Weed, Dillisk	6 *Diodora graeca* Keyhole Limpet
3 *Alaria esculenta* Dabberlocks	7 *Actinia equina* Beadlet Anemone
4 *Cancer pagurus* Edible Crab	8 *Blennius ocellaris* Butterfly Blenny

but can be seen to be ideally adapted to survive these forces. It is essential that they can because once a seaweed becomes detached from the rock it is effectively dead. The holdfast is a most effective anchor, and the great flexibility of the stipe and fronds (which allows the plant to collapse in air) is ideal for absorbing the energy contained in the waves.

On shores where there is a heavy cover of brown seaweeds, with underlying red seaweeds, the change in character between low and high tide is quite remarkable. The large brown seaweeds collapse when exposed to air, covering the rocks. In water, however, they are buoyant, many have air bladders (e.g. bladder wrack) which help them to stand upright and spread out in the water. They create a forest-like canopy, under which there is a mosaic of smaller seaweeds and areas of bare rock, over which the many intertidal animals roam. What appears like just so many layers of plant and animal life piled on top of each other at low tide, becomes an active, highly structured community at high tide.

When covered by water, the intertidal animals are stimulated into action and quickly begin to search for food. The ubiquitous limpet leaves its home scar to graze the surfaces of nearby rocks. However, it always takes care to signpost the way back so that it can safely return to its home scar and the protection it provides.

Barnacles are cemented to the rock and are unable to move about. When covered by seawater they open the central plates of their shells to reveal six pairs of legs. These protrude as a feathery halo, beating furiously, creating a current which draws microscopic plankton towards the mouth of the imprisoned barnacle.

The sea anemone, like the barnacle, cannot go in search of its prey but waits for its prey to be brought to it. It extends its crown of tentacles to capture planktonic animals in the water by firing poisoned darts. Other predators must go looking for their next meal. One of the most common predators on rocky shores is the dogwhelk, which attacks barnacles and mussels. It gains access to its prey either through the open plates on the top of the barnacle shell or by chemically boring a hole in the shell of mussels or other molluscs. The unfortunate victim is digested in its own shell before being sucked up by the dogwhelk.

Many species of fish shelter under seaweeds or stones in any moist area when the tide is out. When the tide is in they patrol the shallow waters in search of succulent pieces of soft tissue, such as the extended tentacles of an anemone. Some of the larger fish will take whole prey, such as worms or molluscs.

On sandy and muddy shores where the surface evidence of life is very limited when the tide is out, the true density of certain inhabitants is revealed when the shore is submerged. The animals buried in the sand or mud are stimulated into action by the seawater percolating through the sediment. What appeared a desert becomes an area of very great, yet still discreet, activity. While beneath the surface the predatory common necklace shell burrows through the sand in search of bivalves which it attacks. Movement through the sand is much easier at high tide, when the sand is lubricated by the seawater and many other animals, including the sea potato, take advantage of this.

Green Seaweeds (Chlorophyceae)

The colour of these seaweeds (algae) comes from the green pigment, chlorophyll, that is used for photosynthesis. Green seaweeds are mostly opportunistic species, rapidly colonizing any unoccupied space at all levels on the shore, although they are most commonly found on the upper shore. Usually, they attach themselves to rocks or other seaweeds, but some species are found on sandy or muddy shores, either as a filamentous mats or fixed to stones and shells.

The structure of these plants varies from the simplest, single-celled species, through the more complex tubular forms (e.g. *Enteromorpha intestinalis*) to the frond-like plants such as *Codium tomentosum*. The genus *Enteromorpha* is especially well adapted to tolerate low salinity and relatively high levels of pollutants, so it occurs in places that are unsuitable for most other seaweeds.

Enteromorpha compressa

(*North Sea, Baltic Sea, English Channel, Atlantic.*) To 30cm in length. Tubular or compressed frond, bearing branches and sometimes branchlets, both noticeably constricted at base. Various shades of green. All shore levels.

Enteromorpha intestinalis

(*North Sea, Baltic Sea, English Channel, Atlantic.*) To 1m in length. Small disc-like holdfast. Short cylindrical stipe. Frond unbranched, tubular and irregularly inflated, which is most obvious on sunny days. Green. Favours upper shore rockpools and reduced salinity.

42

Ulva lactuca
Sea Lettuce

(*North Sea, Baltic Sea, English Channel, Atlantic.*) To 50cm in length. Small disc-like holdfast. Stipe short and solid. Frond variably shaped, flat membrane. Often grows in bunches. Translucent green. Middle shore downwards.

Blidingia minima

(*North Sea, English Channel, Atlantic.*) To 10cm in length. Frond unbranched or slightly branched, possibly inflated. Soft and delicate. Green. Upper shore.

Prasiola stipitata

(*North Sea, English Channel, Atlantic.*) To 1.25cm in length. Stipe moderately long with curled edges. Frond oval to oblong with curled edges, tapering to stipe. Dark green. Upper shore.

Monostroma grevillei

(*North Sea, Baltic Sea, English Channel, Atlantic.*) To 35cm in length. Frond soft, delicate and funnel-shaped. Pale green. Lower shore.

Spongomorpha arcta

(*North Sea, Baltic Sea, English Channel, Atlantic.*) To 7cm in length. Frond much branched, branches often entwined and tangled forming dense, matted tufts. Some branches grow downwards, helping with attachment. Dark green. Middle shore downwards.

Green Seaweeds

Derbesia marina

(*English Channel, Atlantic.*) To 5cm in length. Growth of fine filaments with occasional branches. Bright green. Upper shore downwards.

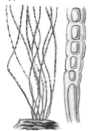

Chaetomorpha linum

(*North Sea, Baltic Sea, English Channel, Atlantic.*) To 30cm in length. Holdfast a single modified cell. Frond cylindrical, fine and rather stiff; cells large enough to see with a hand lens. Often grows in closely packed tufts. Dark green. In pools or covered by sand. Upper to middle shore.

Cladophora rupestris

(*North Sea, Baltic Sea, English Channel, Atlantic.*) To 10cm in length. Frond irregularly and much branched. Appears and feels rather coarse and wiry. Can form extensive carpets. Dark green. Middle shore downwards.

Bryopsis plumosa

(*North Sea, Baltic Sea, English Channel, Atlantic.*) To 10cm in length. Frond glossy, limp, fan-like, with paired branches bearing alternate branchlets, producing a feather-like appearance. Various shades of green. Prefers steep-sided rockpools. Middle shore downwards.

Codium tomentosum

(*North Sea, English Channel, Atlantic.*) To 40cm in length. Holdfast disc-like, made-up of many entangled filaments. Frond tubular, spongy with a felt-like texture, dichotomously branched, typically with a 'brushed' appearance, becoming more shiny with age. Dark green. Middle shore downwards.

Brown Seaweeds (Phaeophyceae)

These are algae in which the green photosynthetic pigment, chlorophyll, is masked by the brown pigment, fucoxanthin. They are often the dominant zone-forming seaweeds on the shore, usually attached to rock.

Ectocarpus siliculosus

(*North Sea, Baltic Sea, English Channel, Atlantic.*) To 50cm in length. Attached to rocks, stones or often epiphytic. Tangled growth form with many irregularly branched, fine hairs, best observed in water. Olive or yellowish-brown. Middle shore downwards.

Ralfsia spp

(*North Sea, English Channel, Atlantic.*) Variable size. Thin, dark brown encrustation. Surface either smooth or warty. On more exposed shores. Middle shore downwards.

Leathesia difformis

(*North Sea, English Channel, Atlantic.*) To 5cm in diameter. Irregularly spherical, shiny growths. More or less solid when young, becoming hollow with age. Yellowish-brown. Often epiphytic on *Laurencia spp* (p.55) and *Corallina spp* (p.52). Middle shore downwards.

Chordaria flagelliformis

(*North Sea, Western Baltic Sea, English Channel, Atlantic.*) To 70cm in length. Disc-like holdfast. Irregularly, much branched frond, often slippery to touch and covered with fine hairs. Dark brown. Usually middle shore downwards.

Dictyota dichotoma

(*North Sea, English Channel, Atlantic.*) To 30cm in length. Disc-like holdfast. Thin, flattened frond, no midrib, regular dichotomous branching. Tips of fronds notched. Fragile appearance. Yellow or olive brown. Middle to lower shore.

Chorda filum

Dead Men's Ropes, Mermaid's Fishing Lines

(*North Sea, Baltic Sea, English Channel, Atlantic.*) To 8m in length. Unbranched, cylindrical slimy frond, made up of air filled compartments. Often tangled with other weeds when washed ashore. Olive brown. Lower shore.

Padina pavonica

Peacock's Tail

(*English Channel, Atlantic.*) To 12cm in length. Narrow stipe producing concave, fan-shaped frond. Chalky deposits on surface. Outer surface brown with greenish bands, inner surface pale green, darkening to stipe. Lower shore.

Alaria esculenta
Dabberlocks

(*North Sea, Northern Atlantic.*) To 50cm in length. Branching holdfast. Cylindrical stipe, possibly bearing bunches of strap-like fruiting bodies. Frond comprising flattened midrib with often torn lamina. Dark brown midrib, yellow-brown lamina. Characteristic of high exposure. Lower shore.

Laminaria digitata
Tangle, Oarweed

(*North Sea, Baltic Sea, English Channel, Atlantic.*) To 2m in length. Large, branching holdfast. Smooth, flexible, oval stipe, gradually broadening into frond. Stipe usually without epiphytes. Orange-brown. Lower shore and rockpools.

Laminaria hyperborea
Cuvie

(*North Sea, Baltic Sea, English Channel, Atlantic.*) To 3.5m in length. Large, branching holdfast. Rough, stiff, round stipe, which tapers away from holdfast. Stipe broadens suddenly into wide digitate frond. Stipe usually with epiphytes. Dark orange-brown. Bottom of lower shore.

Laminaria saccharina
Poor Man's Weather Glass, Sea Belt, Sugar Kelp

(*North Sea, English Channel, Atlantic.*) To 4m in length. Branching, apparently two-tiered holdfast. Relatively short stipe broadening into a crinkly blade/frond with wavy edges. Orange-brown. Favours sheltered areas. Lower shore.

Saccorhiza polyschides

Furbelows

(*North Sea, English Channel, Atlantic.*) To 4m in length. Holdfast characteristic: a large, warty bulb with rhizoid attachments at base. Stipe flattened and twisted near base, with wavy laminae either side. Blade/frond broadens suddenly into wide, feathery lamina divided into straps. Orange-brown. Lower shore.

Ascophyllum nodosum

Egg or Knotted Wrack

(*North Sea, English Channel, Atlantic.*) To 1.5m in length. Disc-shaped holdfast. Short, rounded stipe. Frond flattened, with lightly serrated edge, interrupted at regular intervals by single, large, ovoid air bladders. Golden-yellow reproductive bodies on short stalks. Greenish-brown. On sheltered shores. Middle shore.

Fucus ceranoides

Horned Wrack

(*North Sea, English Channel, Atlantic.*) To 80cm in length. Dichotomously branched frond, with narrow, prominent midrib and relatively thin lamina. Delicate appearance. Terminal, pointed, reproductive bodies. Plant broadly fan-shaped. Yellowish-brown. Middle shore with freshwater influence.

Fucus serratus

Saw or Toothed Wrack

(*North Sea, Baltic Sea, English Channel, Atlantic.*) To 1.8m in length. Branched holdfast. Short stipe. Tough, flattened, branching fronds with distinct midribs, serrated margins and fruiting bodies on end. Often tufts of minute white hairs on fronds. Orange-brown. Middle to lower shore.

Fucus vesiculosus

Bladder Wrack

(*North Sea, Baltic Sea, English Channel, Atlantic.*) To 90cm in length. Disc-like holdfast. Short stipe. Dichotomously branched frond with entire, and often wavy, edge and distinct midribs. Bears paired air bladders either side of midrib. Yellowish reproductive bodies at tips of fronds. Olive-brown. Prefers sheltered shores. Middle shore.

Fucus spiralis

Flat or Spiral Wrack

(*North Sea, English Channel, Atlantic.*) To 50cm in length. Broad, branching frond, with entire edge, conspicuous midribs, no bladders and somewhat spirally twisted, particularly obvious when held by base. Terminal, swollen reproductive bodies. Yellow-brown. Top of shore.

Pelvetia canaliculata

Channel Wrack

(*North Sea, English Channel, Atlantic.*) To 15cm in length. Fronds branched, edges turned-up, forming a distinct channel. No midrib. Tips of fronds bear swollen, pointed fruiting bodies. Grows in dense tufts. Dries out to become brittle, black and shrivelled – easily rehydrated. Olive-green. Upper shore.

Brown Seaweeds

Himanthalia elongata

Thongweed

(*North Sea, English Channel, Atlantic.*) To 2m in length. Insignificant holdfast. Button-shaped frond. Strap-like, dichotomously branching reproductive processes arising from centre of frond. Frond, olive-brown; straps, yellowish-brown. Prefers exposed sites. Lower shore.

Bifurcaria bifurcata

(*English Channel, Northern Atlantic.*) To 50cm in length. Branched, expansive holdfast. Frond cylindrical, smooth and dichotomously branched, with few small bladders. Frond tips bear swollen, oval, pointed fruiting bodies. Olive. Confined to rockpools.

Halidrys siliquosa

Podweed, Sea Oak

(*North Sea, English Channel, Atlantic.*) To 1.5m in length. Small, disc-like holdfast. Frond compressed, alternately branched producing zigzag effect. Long, pointed, pod-like, internally divided air bladders at ends of branches. Ginger-brown. Rockpools and bottom of lower shore.

Sargassum muticum

Japweed

(*English Channel – spreading.*) To 4m in length. Frond much branched, bearing spear-shaped 'leaves' and small, stalked, spherical air bladders. Branched reproductive bodies in clusters. Brown. Lower shore.

Red Seaweeds (*Rhodophyceae*)

In these seaweeds the red pigment phycoerythrin masks the green chlorophyll. Although certain red seaweeds show zonation, their distribution on the shore is mainly determined by local conditions; most are to be found in crevices, under other seaweeds or in rockpools, attached to rock or boulders. However, the diversity of red seaweeds is greater on the lower shore. There is a great variety in both form and structure.

Porphyra umbilicalis

Laver

(*North Sea, Atlantic.*) To 25cm in length. Holdfast producing irregularly shaped, thin, membranous fronds. Polythene-like texture when rubbed together. Purplish-red to olive, darkens when dry, easily bleached. All shore levels.

Gelidium latifolium

(*North Sea, English Channel, Atlantic.*) To 20cm in length. Small holdfast. Frond flat, ribbon-like, with irregular side branches. Frond and branches bear numerous, fine branchlets. Various shades of red. Lower shore.

Palmaria palmata

Dulse, Sheep's Weed, Dillisk

(*North Sea, English Channel, Atlantic.*) To 30cm in length. Disc-like holdfast. Frond roughly wedge-shaped, flattened, variable in outline and with small leaflets growing out from margins. Often found on *Laminaria spp* (p.47) stipes. Dark ruby red. Middle shore downwards.

Dilsea carnosa

(*North Sea, English Channel, Atlantic.*) To 30cm in length. Disc-like holdfast, producing number of short rounded stipes, each developing into a tough, leathery, flattened frond, often torn in older specimens. No leaflets from margin. Dark red. Lower shore.

Red Seaweeds

Hildenbrandia rubra

(*North Sea, Baltic Sea, English Channel, Atlantic.*) Encrusting patches. Often in crevices, runnels or under brown seaweeds. Reddish-brown. Middle shore downwards.

Lithophyllum incrustans

(*North Sea, English Channel, Atlantic.*) Extensive, encrusting patches. Typically has a thick, smooth or bumpy crust, with chalky texture when dry. High ridges form where adjacent patches meet. Pale greyish-pink to mauve. On more exposed shores. Middle shore downwards.

Dumontia contorta

(*North Sea, English Channel, Atlantic.*) To 50cm in length. Disc-like holdfast. Frond tubular with irregular branches, constricted at base and with blunt ends. Maybe twisted or curled. Pale yellow to reddish-purple or brown. Middle shore.

Corallina officinalis

(*North Sea, Baltic Sea, English Channel, Atlantic.*) To 12cm in length. Disc-like holdfast. Stem with oppositely arranged branches and branchlets. Comprising a series of articulated, calcareous segments. Pinkish-red, bleached white by sun. Middle shore.

Lithothamnion spp

(*North Sea, Baltic Sea, English Channel, Atlantic.*) Extensive encrusting patches. Difficult to distinguish from *L. incrustans* in the field, except when it forms nodules. Various shades of pink and mauve. Lower shore.

Ahnfeltia plicata

(*North Sea, English Channel, Atlantic.*) To 15cm in length. Disc-like holdfast. Frond round, much branched, forming stiff wiry tufts. Almost black. Prefers more exposed shores. Middle shore downwards.

Mastocarpus stellatus

Carrageen

(*North Sea, English Channel, Atlantic.*) To 20cm in length. Frond dichotomously branched; margins turned inwards to form channel. When mature, pimple-like fruiting bodies in channel. Dark reddish-brown. Middle shore downwards.

Chondrus crispus

Irish Moss, Carragheen

(*North Sea, Baltic Sea, English Channel, Atlantic.*) To 15cm in length. Frond dichotomously branched, flat, varying widths. Iridescent in water. Dark reddish-brown to purple. Middle shore downwards.

Furcellaria lumbricalis

(*North Sea, Baltic Sea, English Channel, Atlantic.*) To 25cm in length. Claw-like holdfast. Frond cylindrical, regularly branched, stiff, shiny, pod-like reproductive bodies on ends in summer. Reddish-brown. Lower shore.

Cystoclonium purpureum

(*North Sea, Baltic Sea, English Channel, Atlantic.*) To 60cm in length. Root-like holdfast. Frond profusely and irregularly branched; branches constricted at both ends. Reproductive bodies as swellings near tips. Purplish-red. Middle shore downwards.

Lomentaria clavellosa

(*North Sea, English Channel, Atlantic.*) To 40cm in length. Main stem undivided, producing opposite or alternate branches, bearing branchlets which taper at ends. Tends to lie flat in one plane. Bright pinkish-red. Lower shore.

Plumaria elegans

(*North Sea, English Channel, Atlantic.*) To 10cm in length. Small, fibrous holdfast. Frond alternately branched, branches of irregular length, with branchlets alternate or opposite. Plant soft and flaccid. Purplish or brownish-red. Lower shore.

Lomentaria articulata

(*North Sea, English Channel, Atlantic.*) To 25cm in length. Insignificant holdfast. Cylindrical frond, but regularly constricted to appear like strings of beads. Primary branching is dichotomous, secondary branching may be opposite. Purple to bright red. Upper/middle shore downwards.

Ceramium rubrum

(*North Sea, Baltic Sea, English Channel, Atlantic.*) To 30cm in length. Tiny, cone-like holdfast. Irregular dichotomous branching of frond, with pincer-like ends. Banded appearance. Deep red to reddish-brown. All shore levels.

Apoglossum ruscifolium

(*North Sea, English Channel, Atlantic.*) To 10cm in length. Frond broad, leafy, alternately branched, with midrib, wavy edges and rounded tips. Easily torn. Red to pale pink. Middle shore downwards.

Deleseria sanguinea

Sea Beech

(*North Sea, Baltic Sea, English Channel, Atlantic.*) To 40cm in length. Stipe thick and branched, producing blade-like fronds with ruffled, but not indented, edges. Conspicuous midrib and opposite veins. Deep pink. Middle/lower shore.

Hypoglossum woodwardii

(*North Sea, English Channel, Atlantic.*) To 20cm in length. Disc-like holdfast. Several leaf-like, narrow fronds, with pointed tips, conspicuous midribs and branches arising alternately from midribs. Rose to pale pink. Middle shore downwards.

Membranoptera alata

(*North Sea, Baltic Sea, English Channel, Atlantic.*) To 20cm in length. Disc-like holdfast. Frond with irregular dichotomous branches, conspicuous midrib and often notched tips. Various shades of red. Middle shore downwards.

Laurencia obtusa

(*North Sea, English Channel, Atlantic.*) To 15cm in length. Small disc-like holdfast with rootlets. Main stem with opposite branches and branchlets, sometimes arranged spirally and decreasing in size to apex. Red, pink, purple or yellowish. Lower shore.

Laurencia pinnatifida

Pepper Dulse

(*North Sea, English Channel, Atlantic.*) To 20cm in length. Disc-like holdfast with rootlets. Frond thick, fleshy, alternately branched. Purplish-brown to greenish-yellow. Middle shore downwards.

Lichens

Lichens are a seemingly, mutually beneficial partnership between a fungus and an alga. In most species, the shape and form of the main part of the lichen, called the thallus, is determined by the fungal partner. It may be soft or brittle, tufted or encrusting, rough or smooth. Field identification is dependant on the form and colour of the thallus, together with the structure and shape of the fruiting bodies. All the species in this book can be relatively easily identified by eye. Most lichens are found on the upper shore or in the splash-zone, where they are dominant. On more exposed shores they often create spectacular microforests, with their diverse shapes and contrasting colours.

Anaptychia fusca

(*North Sea, English Channel, Atlantic.*) Thallus flat to rock, with close-set lobes. Black, roughly round fruiting bodies with crenulate margins. Thallus golden to dark brown, brown-green when wet.

Arthopyrenia halodytes

(*North Sea, English Channel, Atlantic.*) Thallus found in minute cavities in shells of limpets and barnacles. Dark brown.

Calloplaca marina

(*North Sea, English Channel, Atlantic.*) To 10cm across. Thallus flattish, consisting of small, coarse, granular lobes. Roughly round, convex fruiting bodies. Thallus yellow-orange to bright orange; fruiting bodies deep red-orange.

Lecanora atra

(*North Sea, English Channel, Atlantic.*) To 8cm across. Thallus encrusting, smooth or warty. Fruiting bodies concentrated towards the centre of thallus, with black centre and thick margin. Thallus light to medium grey.

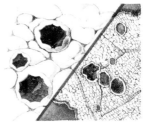

Lichina pygmaea

(*North Sea, English Channel, Atlantic.*) Thallus to 1cm high. Branched tufts form extensive mats. Similar to a minute seaweed in appearance. Dark brown to black.

Ochrolechia parella

(*North Sea, English Channel, Atlantic.*) To 10cm across. Thallus encrusting, thick and warty, forming large, roughly round patches. Fruiting bodies abundant, light grey to white centres, rough, with thick margins. Thallus buff–grey with white marginal area.

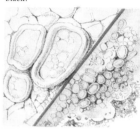

Ramalina siliquosa
Sea Ivory

(*North Sea, English Channel, Atlantic.*) To 5cm across. Thallus erect or hanging down, brittle, glossy or warty. Disc-like fruiting bodies near tip of thallus. Thallus yellowish-grey to greenish-grey; fruiting bodies pale fawn.

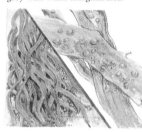

Verrucaria maura

(*North Sea, English Channel, Atlantic.*) Extensive, thick, encrusting thallus. Cracked surface visible with hand lens. Black. Not to be confused with dried oil slick.

Xanthoria parietina

(*North Sea, English Channel, Atlantic.*) Thallus with wrinkled lobes and turned-up edges. Centre of thallus dies out in old specimens. Fruiting bodies toward centre of thallus, orange with pale margin. Thallus bright orange.

Sponges (Porifera)

North European sponges are generally rather small and certainly bear no resemblance to the typical bathroom sponge. Although sponges are simple, sessile (stationary) animals they are extremely difficult to identify in the field, often requiring microscopic examination of the minute spicules, which form the skeleton of the sponge. Basic field identification is based on a combination of growth form, colour, texture and habitat; form and colour can be very variable and therefore not reliable in themselves. Sponges feed by drawing in water through many small, often inconspicuous holes (ostia) and then extracting suitable food particles, before forcing the water out through a large, often conspicuous hole (osculum) at the top.

Clathrina coriacea

(*North Sea, English Channel, Atlantic.*) Encrusting, formed from many branched, twisted tubes. A number of tubes join at a single osculum. Maybe quite extensive and thick. Grey, yellow, brown or red. Lower shore. (Number of similar species in genus.)

Grantia compressa

Purse Sponge

(*North Sea, English Channel, Atlantic.*) To 5cm high. Roughly vase-shaped, flattened with large osculum at apex. Attached by a short stalk, often hanging in clumps on underside of overhangs. Grey, beige or pale yellow. Middle shore downwards.

Scypha ciliata

(*North Sea, English Channel, Atlantic.*) To 5cm high. Cylindrical, vase-shaped, erect, borne on short stalk. Overall rough appearance, with large apical osculum surrounded by fringe of stiff spicules. Green or yellow. Lower shore.

Cliona celata

Boring Sponge

(*North Sea, Baltic Sea, English Channel, Atlantic.*) Bores into limestone and shell. System of channels makes shells fragile. On outer surface, many tiny (less than 2mm in diameter) holes; yellow or green visible inside.

Hymeniacedon sanguinea

(*North Sea, Baltic Sea, English Channel, Atlantic.*) Encrusting, forming irregular growths. Surface furrowed, ridged or smooth, with small random oscula. Orange, scarlet or deep-red. Middle shore but becoming more abundant on lower shore.

Oscarella lobularis

(*North Sea, English Channel, Atlantic.*) Encrusting, forming patches of irregular shape and varying size and thickness, usually lobed. Often brightly coloured, variously pink, yellow, brown or violet. Middle shore downwards.

Halichondria panicea

Breadcrumb Sponge

(*North Sea, Baltic Sea, English Channel, Atlantic.*) Encrusting, forming thick and extensive patches. Surface relatively smooth, oscula often borne on raised areas. Green most common colour, but also yellow, orange, brown or white. Middle shore downwards.

Myxilla incrustans

(*North Sea, English Channel, Atlantic.*) Encrusting, forming large, cushion-like clumps. Wrinkled surface with large oscula, unevenly arranged. Typically orange but may be brown or yellow. Often found attached to stones on muddy sand. Lower shore.

Sea Firs, Jellyfish, Sea Anemones and Corals (Cnidaria)

Members of the phylum Cnidaria are amongst the most common of the marine animals. They have a very simple, basic construction, but are highly variable in their appearance. The phylum contains three classes; the Hydrozoa (sea firs) and Anthozoa (sea anemones and corals) are both abundant on the shore, while the Scyphozoa (jellyfish) are essentially animals of the open seas, but are often found stranded on the shore.

Sea firs are microscopic colonial animals which attach themselves to rocks or seaweeds. Most individuals, or polyps, within a colony are specialized for feeding, while others are primarily concerned with defence or reproduction.

Jellyfish consist of a saucer-shaped jelly mass with a central mouth on the underside, surrounded by long, trailing tentacles with which they capture their prey. Once stranded they die and rapidly decompose, making species identification difficult.

Sea anemones and corals are sedentary, soft-bodied animals usually found anchored to a rock or buried in sand or mud. Sea anemones are territorial, aggressive, colonial or solitary, and are much more common on the shore than corals. They are easily spotted because of their bright colouring.

There are two types of corals; soft corals, which are always colonial, and hard corals, which are either solitary or colonial. Each individual animal is known as a polyp. Hard coral polyps have an outer, limestone skeleton; the soft corals have a soft, flexible, shared skeleton made up of numerous tiny spicules.

Cerianthus lloydii

(*North Sea, English Channel, Atlantic.*) To 15cm in height, tentacles span to 7cm. Large number of slender tentacles, arranged in two concentric rings; typically brown, green, white or banded and retract rapidly if disturbed. Body, dull yellow colour. Embedded in fine sand or mud at low water.

Actinia equina

Beadlet Anemone

(*North Sea, English Channel, Atlantic.*) To 7cm in height, 5cm diameter at base. Tentacles in five or six rings, with 24 blue spots in circle at base. Variable colour, shades of red, green, brown or orange. All shore levels.

Anemonia viridis

Snakelocks Anemone

(*English Channel, Atlantic.*) To 10cm in height, tentacles span to 20cm. Smooth column with many long, wavy tentacles, rarely retracted. Column brownish- or greyish-green, tentacles similar or bright green with purple tips. Prefers bright, sunlit areas. Lower shore.

Alcyonium digitatum

Dead Man's Fingers

(*North Sea, English Channel, Atlantic.*) To 25cm in length. Erect, branching colonies usually divided into a few blunt fingers. Under water has finely hairy appearance, due to extended tentacles. Shades of either white, yellow, pink or orange. Occurs on rock in sheltered locations. All shore levels.

Urticina felina

Dahlia Anemone

(*North Sea, Baltic Sea, English Channel, Atlantic.*) To 15cm in height, tentacles span to 20cm. Covered with sand grains, shell debris and gravel. Tentacles short, stout, in multiples of ten. Column, variable in colour and pattern, with blues, greens and reds. Tentacles similarly banded. Lower shore.

Metridium senile

Plumose Anemone

(*North Sea, English Channel, Atlantic.*) To 10cm in height. Smooth column with prominent collar. Numerous, densely packed, slender tentacles. Always plain and unpatterned. Usually white or orange, sometimes yellow or reddish–pink. Prefers strong water movement. (Middle)/lower shore.

Sagartia elegans

(*North Sea, English Channel, Atlantic.*) To 6cm in height, tentacles span to 7cm. Column bearing suckers, appearing as pale spots. Tentacles numerous in five or six circles. Many colour varieties; red, brown, orange, green and white. Tentacles often contrasting, maybe banded. Middle shore downwards.

Cereus pedunculatus
Daisy Anemone

(*English Channel, Atlantic.*) To 7cm in height. Column smooth and trumpet-shaped, bearing grey suckers with adhering gravel. Many short tentacles. Column brownish, sometimes yellowish or reddish, and flecked. Tentacles often brown with cream patterning. Middle shore downwards.

Peachia hastata

(*North Sea, English Channel, Atlantic.*) To 30cm in height, tentacles span to 12cm. Column worm-like with rounded base, maybe covered with adhering sand. 12 tentacles in single ring. Column flesh-colour, pink or buff with longitudinal lines. Tentacles highly patterned, cream and brown. Lower shore.

Halcampa chrysanthellum

(*North Sea, Baltic Sea, English Channel, Atlantic.*) To 7cm high. Column elongated and worm-like, with bulbous base. 12 short tentacles. Column, white, pink or buff, longitudinally striped, alternate opaque and translucent. Tentacles with reddish-brown banding. Bottom of lower shore.

Edwardsia claperedii
Worm Anemone

(*North Sea, English Channel, Atlantic.*) To 10cm in height, tentacles span to 5cm. Column elongated and worm-like. 16 very long, delicate tentacles in two circles. Column translucent pink, tentacles transparent with white and brown dots. Lower shore.

Bunodactis verrucosa
Gem Anemone

(*English Channel, Atlantic.*) To 3cm in height, tentacles span to 6cm. Column bearing six, longitudinal rows of white warts. Tentacles in circles of six, up to 48 in total. Column pink or grey with mottling of greens, reds and browns near base. Tentacles pink or grey but banded white. Often in bright sunlit areas. Lower shore.

Corynactis viridis
Jewel Anemone

(*English Channel, Atlantic.*) To 1.5cm in height. Small, smooth columned anemone, with numerous tentacles, each ending in a small knob. Often brilliantly coloured in green, pink, red, orange, brown and white. Knobs often have a contrasting colour. Common in large groups on rocks in shade. Lower shore.

Caryophyllia smithii
Devonshire Cup-coral

(*North Sea, English Channel, Atlantic.*) To 1.5cm in height. Stony coral with conspicuous ridges on outer surface of cup. Tentacles with spherical knob at end. Colour highly variable, orange, red, brown, green, white, pink with contrasting pattern round rim. Bottom of lower shore.

Ribbon Worms (Nemertina)

Nemertines are fragile, unsegmented worms with smooth, slimy bodies. They move in a characteristic way, by passing waves of contractions along their length. Ribbon worms are carnivorous, generally found in sand or mud burrows, or under boulders. They are relatively difficult to identify in the field – not least because their bodies are highly contractile, which means that their size is extremely variable – but the arrangement of eyes is a useful feature.

Lineus longissimus

Bootlace Worm

(*North Sea, Baltic Sea, English Channel, Atlantic.*) To 5m or more in length. Head bears numerous, inconspicuous eyes. Very long, tubular body with constriction behind head, tapering abruptly at tail. Dark brown or black with purplish sheen. Lower shore.

Cephalothrix rufifrons

(*North Sea, English Channel, Atlantic.*) To 10cm long. No eyes. Body generally tapers towards both ends. Pale yellow or flesh colour, reddish towards head, dark line down centre of body. Lower shore.

Lineus ruber

Red Ribbon Worm

(*North Sea, Baltic Sea, English Channel, Atlantic.*) To 20cm in length. Row of three or four eyes on either side of spatula-like head, which tapers towards body. Body tapers towards tail. Reddish-brown, dorsal surface darker than ventral. Middle shore downwards.

Amphiporus lactifloreus

(*North Sea, Baltic Sea, English Channel, Atlantic.*) To 8cm in length. Spatula-like head, with four groups of eyes: a row each side of head; a group each side of a ganglion, which appears as a pink patch on head. Ventrally flattened body. Various shades of white to pink, with translucent line down back. Often associated with Laminarians (p.47). Lower shore.

Segmented or Bristle Worms (Annelida)

The majority of marine annelids belong to the class Polychaeta. The other two classes of annelids, which include such animals as leeches and earthworms, are mostly confined to land or freshwater.

The body of an annelid is divided into many small segments, which may bear bristles, known as chaetae. Most species have a well developed head region with eyes, various tentacles and a modified mouth. There are a very large number of species, which can easily be divided into two groups, free-living and tube-dwelling. The tube-dwelling animals can be further divided, depending on the type of tube (mucus, sand or calcareous). The free-living species are generally carnivorous, whilst the tube-dwellers are either herbivores or feed on decaying organic matter on the seabed.

Aphrodite aculeata

Sea Mouse

(*North Sea, Baltic Sea, English Channel, Atlantic.*) To 20cm in length. Body, dorsally convex, ventrally flattened. Covered in green and greyish-brown iridescent hair. Ventral surface covered in golden-brown chaetae. Found in shallow water. Bottom of lower shore.

Harmothoe impar

(*North Sea, Baltic Sea, English Channel, Atlantic.*) To 2.5cm in length. Oval body, slightly pointed at tail. Bears 15 pairs of overlapping scales covering 35-40 segments, bearing chaetae. Fragile, easily damaged when handled. Scales brownish-green with central yellow patch. Middle shore downwards.

Lepidonotus clava

(*English Channel, Atlantic.*) To 3cm in length. Flat body with linear sides and bluntly rounded head and tail. Covered by smooth, rounded, non-overlapping scales, which partly cover the head. Brown with white speckling. Lower shore.

Segmented or Bristle Worms

Anaitides maculata

(*North Sea, Baltic Sea, English Channel, Atlantic.*) To 10cm in length. Body with about 250 segments bearing paddles. Heart-shaped head with pair of large eyes, two pairs of short antennae and four pairs of tentacles. Body yellowish-green with variable pattern of brown spots and bars. Paddles, brown. Lower shore.

Glyceria convoluta

(*North Sea, English Channel, Atlantic.*) To 10cm in length. Body rounded, tapering at both ends, 120-180 segments, each with two rings and small tufts of bristles. Small head with line of four minute antennae. Pair of paddles at anal end. Active and coils up when disturbed. Pink. Lower shore.

Polynoe scolopendrina

(*North Sea, English Channel, Atlantic.*) To 12cm in length. Body elongated. Front end covered by 15 pairs of scales, back end naked. Scales iridescent, greyish-green with red body visible beneath. Naked segments with central brown line and two lateral spots. Lower shore.

Eulalia viridis
Green Leaf Worm

(*North Sea, Baltic Sea, English Channel, Atlantic.*) To 12cm in length. Body with about 200 segments bearing paddles. Head with a pair of black eyes, two pairs of short antennae, a single central antenna and four pairs of tentacles. Body dark green, paler towards head. On rock or *Laminaria spp.* Middle shore downwards.

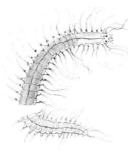

Kefersteinia cirrata

(*North Sea, Baltic Sea, English Channel, Atlantic.*) To 7.5cm in length. Head bears four eyes, two short antennae and eight pairs of tentacles. Body, 50-65 segments bearing chaetae, long white dorsal cirri and short ventral cirri. Very fragile, breaks easily. Red, purple, brown or yellow with distinctive dorsal blood vessel. Lower shore.

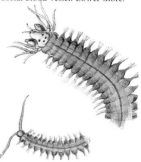

Hediste diversicolor

Rag Worm

(*North Sea, Baltic Sea, English Channel, Atlantic.*) To 12cm in length. Heart-shaped head, four eyes, one pair of short antennae, four pairs of tentacles. Body, 90-120 segments with parapodia bearing chaetae. Variable colour, greenish, yellow, red, orange, with distinct red line down dorsal surface. Middle shore downwards.

Neanthes virens

King Rag Worm

(*North Sea, Baltic Sea, Atlantic.*) To 40cm in length. Head, two very short antennae, four eyes, four pairs of tentacles. Body can be as thick as a finger, 100-175 segments with complex parapodia with chaetae and lobes that make them appear like paddles. Green with purple tints. Parapodia, green with yellow margin. Lower shore.

Perinereis cultifera

(*North Sea, English Channel, Atlantic.*) To 25cm in length. Head, two antennae, four eyes, four pairs of tentacles. Body, flattened and tapering towards back, 100-125 segments bearing parapodia and chaetae. Body brownish-green, prominent dorsal blood vessel. Parapodia, reddish. Lower shore.

Nephtys caeca

(*North Sea, Baltic Sea, English Channel, Atlantic.*) To 25cm in length. Head, four short antennae, no eyes apparent. Body, 90-150 segments each with a pair of large, rounded, bilobed parapodia with long, soft chaetae. Thread-like tail. Body, yellowish-grey with pink and red tints and yellow chaetae. Characteristic rapid wriggling motion. Middle shore downwards.

Scoloplos armiger

(*North Sea, Baltic Sea, English Channel, Atlantic.*) To 15cm in length. Head without any visible appendages or eyes. Body, about 200 segments with front 20 fat and flattened. Towards the back end chaetae are long and point over back. Two threads for tail. Red, orange or brown. Middle shore downwards.

Polydora ciliata

(*North Sea, Baltic Sea, English Channel, Atlantic.*) To 3cm in length. Head, four small eyes in a square and two long, often coiled antennae. Body, 60-180 segments bearing chaetae. Bores into shell, lives in u-shaped tube. Semi-transparent, brownish-yellow.

Cirratulus cirratus

(*North Sea, English Channel, Atlantic.*) To 12cm in length. Head with four to eight pairs of small, black eyes in rows on either side. Body, up to 130 segments, fourth or fifth segment with two bunches of long tentacles. Red thread-like gills on all chaetae-bearing segments. Yellowish-orange, red or brown. Lower shore.

Cirriformia tentaculata

(*North Sea, English Channel, Atlantic.*) To 20cm in length. Head, pointed, no eyes. Body, more than 300 segments with bunches of coiled, filamentous, tentacles on segment six or seven. Thread-like gills on all but the last few chaetae-bearing segments. Yellowish-orange, brown or bronze, tentacles and gills red. Middle shore downwards.

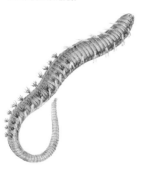

Capitella capitata

(*North Sea, Baltic Sea, English Channel, Atlantic.*) To 10cm in length. Head with two very small black eyes. Body tapers markedly to both ends, up to 100 segments with chaetae. Blood red. Middle shore downwards.

Arenicola marina
Lugworm

(*North Sea, Baltic Sea, English Channel, Atlantic.*) To 20cm in length. Body in two distinct parts: front 19 segments, fat, soft and shiny with chaetae, segments 7-19 bear gills; back section, narrower, stiffer, no chaetae or gills. Greenish-black, red gills. Lives in u-shaped tube, produces characteristic sand cast. Middle shore downwards.

Owenia fusiformis

(*North Sea, English Channel, Atlantic.*) To 10cm in length. Head with six branched gills. Body slender, 20-30 segments of varying lengths, all with chaetae. Greenish-yellow. Lives in mucus tube covered in sand, part of which protrudes above surface. Lower shore.

Amphitrite gracilis

(*North Sea, English Channel, Atlantic.*) To 12cm in length. Body, long and gelatinous, 100-200 segments, two pairs of branched gills, numerous long tentacles, 17-19 of front segments bear chaetae. Pale red or yellow, gills red, tentacles pink. Mucus burrow on lower shore.

Sabella pavonina

Peacock Worm

(*North Sea, English Channel, Atlantic.*) To 25cm in length. Head, complete crown of gills of up to 90 filaments. Body, 100-600 small segments, 6-12 segments with chaetae. Pale greenish-grey, orange and violet tints; gills brown, red, violet, with or without banding. Flexible tube. Lower shore.

Sabellaria alveolata

Honeycomb Worm

(*North Sea, English Channel, Atlantic.*) To 4cm in length. Head with three concentric rings of chaetae, which form stopper for tube. Body, 32-37 segments with sickle-shaped gills, except on slender, terminal section. Red or brownish-red, purplish head. Colonial, many tubes of large sand grains give appearance of honeycomb. Lower shore.

Lanice conchilega

Sand Mason Worm

(*North Sea, English Channel, Atlantic.*) To 30cm in length. Head with palps. Body, long, thin; 150-300 segments, two to four bear pairs of gills and tentacles, front 17 bear chaetae. Reddish-pink, yellow or brown; red gills and near-white tentacles. Stiff tube. Middle shore downwards.

Pomatoceros triqueter

Keel Worm

(*North Sea, English Channel, Atlantic.*) To 2.5cm in length. Head, two gills each of up to 20 filaments. Body, 80-100 segments bearing chaetae. Reddish-brown. Lives in triangular, calcareous tube with sharp dorsal keel, encrusting stones. Opening at mouth end where gill filaments protrude, tapering to slightly coiled back end. Lower shore.

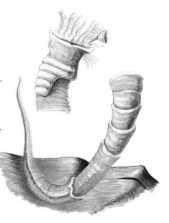

Serpula vermicularis

(*North Sea, English Channel, Atlantic.*) To 7cm in length. Head with two gills, each of up to 30 fine filaments. Body, 200 segments bearing chaetae. Red or yellow. Lives in calcareous, round tube, showing growth lines, usually off-white colour. Fixed to substrate at back end; front end often raised off rock. Tubes tangled together. Lower shore.

Spirorbis borealis

(*North Sea, English Channel, Atlantic.*) To 35mm in length. Head, two gills each of four or five filaments. Body, 21-35 segments bearing chaetae, brown front end, red back end. Gills, dull yellow with greenish tinge. Lives in clockwise-coiled calcareous, off-white tube. Usually no more than two coils. Middle shore downwards. (A number of closely related species.)

Other Worms (Priapuloidea, Sipunculoidea, Echiuroidea)

The priapuloids are cylindrical worms, with bulbous bodies, a mouth at the front end and an anus at the back. They appear segmented, but internally they are no divisions. The front end can be turned inside out and often bears spines. They are carnivorous, living in muddy sand, often under boulders.

The sipunculoids are also cylindrical worms, with thick skins and both mouth and anus located at the front. The mouth is on the end of a proboscis and surrounded by a frill of tentacles. The anus is situated on the top at the front. The sipunculoids often have a chequered appearance due to bands of transverse and longitudinal muscles which can be seen through the skin.

The echiuroids are variable in shape, but generally roughly sausage shaped, with a proboscis of varying length. The mouth is on the underside and is associated with the retractable proboscis on the top side. The anus is at the back end. Some species possess bristles and/or hooks. The female is large, while the male is small and parasitic on the female.

Priapulus caudatus

(*North Sea, Baltic Sea, Atlantic.*) To 8cm in length. Cylindrical body appearing segmented. Front portion bears lines of spines with mouth at end, back end bears a single, branched appendage. Flesh coloured. Lower shore.

Sipunculus nudus

(*North Sea, Atlantic.*) To 20cm in length. Cylindrical body, tough skin with rectangularly chequered surface texture. Papillae on proboscis, mouth surrounded by four tentacles. Yellowish, flesh coloured. Middle shore downwards.

Echiurus echiurus

(*North Sea, English Channel, Atlantic.*) To 15cm in length. Body cylindrical, with rows of small papillae. Front end bears proboscis up to five centimetres long and with two long bristles. Two rows of bristles around anus. Yellow or orange. Lower shore.

Thallasema thallasenum

(*English Channel, Atlantic.*) To 7cm in length. Body tapering at both ends, highly contractile, often slimy. Pointed proboscis with groove and frilled edge. Front bluish or yellowish, middle grey or pink, back opaque white. Lower shore.

Golfingia elongata

(*North Sea, English Channel, Atlantic.*) To 9cm in length. Cylindrical body, smooth, with no papillae, tapering to both ends, highly contractile. Mouth on end of proboscis, surrounded by a number of short, branched tentacles. Pale straw colour, darker brown towards ends. Lower shore.

Bonellia viridis

(*English Channel, Atlantic.*) Female to 15cm in length, male to 2mm. Front end bears a very long proboscis, ending in a forked tip with leaf-like appearance. Groove runs from tip to mouth. Greenish. Found in holes in rock. Bottom of lower shore.

Barnacles, Sandhoppers, Shrimps, Lobsters and Crabs (Crustacea)

There are more crustaceans in the sea than any other class of animals. Like all the other members of the phylum Arthropoda, they are all bilaterally symmetrical, with a segmented body divided into three regions – head, thorax and abdomen. The head is further divided into segments, which bear a variety of appendages. Arthropods have an external skeleton, which cannot be expanded so it has to be moulted periodically. The Crustacea are the most primitive of arthropods. They all have two pairs of antennae, many appendages attached to the segments of head, thorax and abdomen, and gills. The head and thoracic regions are difficult to distinguish, whereas the abdomen is easily recognized.

The barnacles have a typical crustacean, planktonic larval stage in their life cycle. When selecting a suitable site to colonize, the larvae are able to sense chemicals on the rock surface left by earlier generations of barnacles. If these chemicals are present then the larva settles and grows. The adult barnacle, unlike other crustacea, is sessile (stationary); it lives enclosed within a calcium carbonate shell, attached to the rock by the back of its head. Lying in this position, it beats its thoracic legs to create a feeding current, pulling water into the shell and then filtering out the suspended food.

Sandhoppers, shrimps, crabs and lobsters all have a similar body plan, but varying forms. All are highly mobile, carnivorous or scavengers, searching for food using highly developed eyes and/or chemical receptors. Many of their numerous appendages are divided in two, with each branch carrying out a different function. In most cases one branch is used for locomotion and the other respiration.

Lepas anatifera

Goose Barnacle

(*North Sea, English Channel, Atlantic.*) Does not normally live on shore but may be stranded on drift wood or found attached to the hulls of boats. To 5cm shell length, 20cm stalk length. Shell of five plates, almost white with translucent, bluish sheen. Stalk, tubular, retractible with a concertina-like appearance, dark blue-grey.

Verruca stroemia

(*North Sea, English Channel, Atlantic.*) To 6mm in diameter. Shell of four, longitudinally ribbed plates of different sizes. Aperture divided asymmetrically into two halves by central line. Never crowded together. White or brownish-white. Lower shore.

74

Chthamalus stellatus

Star Barnacle

(*North Sea, English Channel, Atlantic.*) To 1cm in diameter. Shell of six fluted plates. Four side plates overlap the end plates. Aperture kite-shaped. White or greyish. Most abundant on exposed shores. Upper and middle shore.

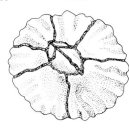

Semibalanus balanoides

Acorn Barnacle

(*North Sea, Baltic Sea, English Channel, Atlantic.*) To 1.3cm in diameter. Shell of six ridged plates. Aperture diamond-shaped. White or greyish. Upper shore downwards. (Where *C. stellatus* present, middle shore downwards.)

Balanus crenatus

(*North Sea, Baltic Sea, English Channel, Atlantic.*) To 2cm in diameter. Shell of six ridged plates. Conical, nearly as tall as wide and appears to lean to one side. Aperture elliptical, toothed, with a yellow rim. Pale grey. Lower shore.

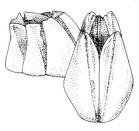

Balanus perforatus

(*English Channel, Atlantic.*) To 3cm in diameter. Nearly as tall as broad. Shell of six smooth, or vertically lined plates which separate at top producing a jagged edge around aperture. Aperture offset. Pale purplish-brown. Lower shore.

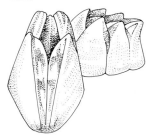

Elminius modestus

Australian Barnacle

(*North Sea, English Channel, Atlantic – range expanding.*) To 1cm in diameter. Shell of four, easily distinguished plates. Each plate with two slight grooves and shell has star-like outline. Pale-grey. Upper to middle shore.

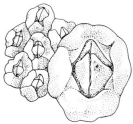

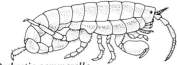

Orchestia gammarella

(*North Sea, Baltic Sea, English Channel, Atlantic.*) To 1.8cm in length. Head with large, almost round, black eyes. Short upper antennae, long lower antennae which are about half length of body. Second thoracic legs end in small claws; third thoracic legs end in large claws. Greenish or reddish-brown. Tide line.

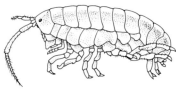

Talitrus saltator

Sandhopper

(*North Sea, Baltic Sea, English Channel, Atlantic.*) To 2.5cm in length. Head with large, black eyes. Short upper antennae, lower antennae two–thirds length of body. Second thoracic legs end in spikes; third end in small claws. Brownish-grey or greenish with black line down back. Tide line.

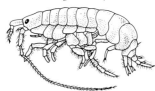

Bathyporeia pelagica

(*North Sea, Baltic Sea, English Channel, Atlantic.*) To 8mm in length. Head with pair of red eyes. Short upper antennae, bend after first segment. Lower antennae as long as body in males; twice the length of upper antennae in females. Colourless. Middle shore.

Gammarus locusta

(*North Sea, Baltic Sea, English Channel, Atlantic.*) Male to 2cm in length, female to 1.4cm. Head with pair of black eyes. Upper antennae branched and longer than lower. Second thoracic appendages end in nippers. First three pairs abdominal appendages for swimming, last three for jumping. Brownish-green with red spots. Middle shore downwards.

Corophium volutator

(*North Sea, Baltic Sea, English Channel, Atlantic.*) To 8mm in length. Head with small, black eyes. Upper antennae half length of body, lower antennae almost twice as long. Last pair thoracic appendages longest. Greyish. Lives in u-shaped burrow, apparent as pin holes in the sediment surface. Middle shore.

Eurydice pulchra

(*North Sea, Baltic Sea, English Channel, Atlantic.*) To 7mm in length. Head with short inner antennae and outer antennae more than half length of body. End segment very large. First three pairs of thoracic appendages short and hooked, remainder long and hairy. Semi-transparent and pale grey. Middle shore downwards.

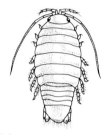

Sphaeroma rugicauda

(*North Sea, Baltic Sea, English Channel, Atlantic.*) To 1cm in length. Body smooth, bluntly oval with seven pairs of similar legs and rough tail section. Inner antennae half length of outer antennae, which are one third length of body. Grey with black streaks and light stripe down back. Swims very actively on back. Curls into ball when disturbed. Favours saltmarsh pools. Upper to middle shore.

Jaera albifrons

(*North Sea, Baltic Sea, English Channel, Atlantic.*) Male to 3mm in length, female to 5mm. Body oval with pronounced indentations between segments, lateral edges ending in spines. Tail section broader than long. Inner antennae very short, outer antennae two thirds length of body. Ash coloured. Upper to middle shore.

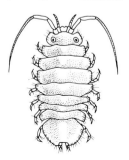

Idotea baltica

(*North Sea, Baltic Sea, English Channel, Atlantic.*) Male to 3.5cm in length, female to 1.7cm. Body oblong with dorsal keel. Tail plate ends in three points, central one longest. Inner antennae short, outer antennae half length of body. Green or brown, maybe with spots or lines. Middle shore downwards.

Ligia oceanica

Sea Slater

(*North Sea, English Channel, Atlantic.*) Male to 2.8cm in length, female to 2cm. Body flattened with rough texture. Striking black eyes. Inner antennae almost regressed, outer antennae two thirds length of body. Pair of forked tail appendages. Upper shore and splash-zone.

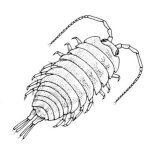

Crangon vulgaris

Common or Brown Shrimp

(*North Sea, Baltic Sea, English Channel, Atlantic.*) To 7cm in length. Small rostrum, carapace with central spine and one either side. Inner antennae in two parts, outer antennae as long as body. First pair of legs with heavy, modified nippers, second pair end in minute pincers. Lower shore.

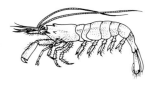

Eupagurus bernhardus

Common Hermit Crab

(*North Sea, Baltic Sea, English Channel, Atlantic.*) To 10cm in length. First pair of legs modified to large, unequal pincers with granulated surface. Second and third pairs of legs end in spines. Red or yellow, brownish abdomen. Inhabits variety of gastropod shells when small, prefers *Buccinum undatum* (p.93) shells when larger. Middle shore downwards.

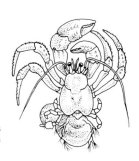

Galathea dispersa

(*North Sea, English Channel, Atlantic.*) To 5cm in length. Flat rostrum, covered with bristles and with central spike and four smaller points either side. Pincers hairy and rough, about same length as body. Red or orange-brown. Lower shore.

Galathea intermedia

(*North Sea, English Channel, Atlantic.*) To 1cm in length. Pointed rostrum with four blunt points either side, near eyes. Pincers very large, approximately twice length of body. Bright red with blue spots. Lower shore.

Galathea squamifera

(*North Sea, English Channel, Atlantic.*) To 5cm in length. Rostrum with three central points of similar size and three smaller points on either side. Pincers very scaly with tubercles and spines on outside of last segment and inside of other segments. Body rough with transverse rows of tubercles. Greenish-brown with red flecks. Lower shore.

Galathea strigosa

(*North Sea, English Channel, Atlantic.*) To 10cm in length. Rostrum sharply pointed with three points either side. Pincers one and a half times as long as body, large, flat and very spiny. First three pairs of walking legs massive and spiny. Red with blue transverse lines. Aggressive when handled. Lower shore.

Psidia longicornis

Long-clawed Porcelain Crab

(*North Sea, English Channel, Atlantic.*) To 6mm in diameter. Shell almost round, not hairy and appears clean and shiny. Three lobes in shell between eyes, middle one deep and grooved. Outside the eyes are a pair of long, fine antennae. Pincers slender. Brown, red or yellow with or without pattern. Lower shore.

Porcellana platycheles

Broad-clawed Porcelain Crab

(*North Sea, English Channel, Atlantic.*) To 1.2cm in diameter. Shell hairy, slightly longer than broad. Three blunt teeth between eyes, a pair of long antennae. Pincer flattened, with hairs on outer edge. Dirty grey-brown or reddish-brown. Middle shore downwards.

Hyas arenarias

Spider Crab

(*North Sea, Baltic Sea, English Channel, Atlantic.*) To 10cm in length. Shell, roughly pear-shaped, many tubercles. Two converging points between the eyes. Walking legs slightly narrower and longer than pincers. Dull reddish-purple, shell often completely covered by seaweeds. Lower shore.

Inachus dorsettensis

Spider Crab

(*North Sea, English Channel, Atlantic.*) To 2.5cm in length. Shell triangular with rounded points and characteristic surface sculpturing. Front of shell with four small bumps and one large bump behind. Rear of shell, single small bump with a large bump either side. Two spines between eyes. Second pair of legs three times body length. Yellowish-brown. Lower shore.

Macropodia rostrata

Spider Crab

(*North Sea, Baltic Sea, English Channel, Atlantic.*) To 1.6cm in length. Shell triangular with eight regularly distributed spines over surface and with a short rostrum. Second and third pairs of legs end in points, fourth and fifth pairs in hooks. Reddish, maybe yellowish-brown. Covered in seaweeds and sponges. Lower shore.

Corystes cassivelanus

Masked Crab

(*North Sea, English Channel, Atlantic.*) To 4cm in length. Shell smooth and not as broad as long. Between eyes, two points and pair of very long, hairy antennae. On sides of shell, three points. First pair of legs have pincers, in males twice body length, in females body length. Other legs hairy. Brownish-yellow. Lower shore.

Pirimela denticulata

(*North Sea, English Channel, Atlantic.*) To 2.5cm in length. Shell, smooth, almost as broad as long. Three points between eyes, outer two triangular, middle one rounded. Seven points either side of shell. Pincers small, all other legs end in points. Green, brown or purple. Lower shore.

Cancer pagurus

Edible Crab

(*North Sea, English Channel, Atlantic.*) To 12cm in length. Shell, granulated texture, roughly elliptical, broader than long. Between eyes, three blunt lobes and a further ten lobes either side round edge of shell. First pair legs with claws, other legs hairy. Pinkish-brown with black tip to claws. Middle shore downwards.

Liocarcinus puber

Velvet Swimming Crab

(*North Sea, English Channel, Atlantic.*) To 8cm in length. Shell with eight to ten small points between pair of red eyes. Five larger points either side of shell. Pincers sometimes unequal. Fifth pair of legs flattened and rounded. Hairy. Reddish-brown with blue joints. Aggressive. Lower shore.

Carcinus maenus

Shore Crab

(*North Sea, Baltic Sea, English Channel, Atlantic.*) To 4cm in length. Shell broad, with three small blunt teeth between eyes and five sharp teeth either side. Fifth pair of legs flattened and pointed. Blackish-green or brown. Middle shore downwards.

Pilumus hirtellus

Hairy Crab

(*North Sea, English Channel, Atlantic.*) To 1.8cm in length. Shell broader than long with deep central fissure between eyes and five small teeth either side. First pair of legs with moderately large, unequal pincers. Hairy. Brownish-red with yellow, pincers brown with dark tips. Lower shore.

Eriochier sinensis

Chinese Mitten Crab

(*North Sea, Baltic Sea, English Channel, Atlantic.*) To 7cm in length. Shell almost square with rounded corners and four points between eyes and three either side. First pair of legs with pincers, the last joints covered in dense hairs. Other legs long and hairy. Olive-green. Middle shore downwards.

Sea Spiders and Insects (Pycnogonida, Insecta)

These 'spiders' are a small group of wholly marine arthropods. They have four pairs of legs, as do the better known terrestrial spiders (Arachnida), but that is where any similarity ends. They are not very common, but always a delight when found. There are a few insects that have adapted to the marine habitat.

Pycnogonum littorale

(*North Sea, English Channel, Atlantic.*) To 1.5cm in length. Body has a prominent projection at the front end and a dorsal tubercle on each segment. Legs appear stumpy and segmented, ending in a claw. Dirty straw colour. Lower shore.

Nymphon gracile

(*North Sea, English Channel, Atlantic.*) To 1cm in length. Body long, slender with very small abdomen region. Bears a pair of pincer-like feeding appendages, pair of five-jointed feelers and four pairs of long legs. Red or pinkish-yellow. Middle shore downwards.

Petrobius maritimus

Bristle-tail

(*North Sea, English Channel, Atlantic.*) To 1.2cm in length. Body ends in hair-like projection, which is the length of the body and flanked by pair of shorter projections. Head with pair of long antennae. Three pairs of thoracic legs. Brown. Upper shore and splash-zone.

Lipura maritima

(*North Sea, English Channel, Atlantic.*) To 3mm in length. Body plump, broadest near rear and then tapering to blunt point. Head with pair of short antennae and three pairs of thoracic legs. Blue-grey. Upper shore and in surface film of rockpools.

Chitons, Snails, Slugs, Bivalves, Squid and Octopus (Mollusca)

Molluscs are bilaterally symmetrical, unsegmented animals with a soft body, which is often enclosed within a hard external skeleton or shell. Although very variable in form, they do have some basic characteristics in common. One of these is that they all possess a foot: the chitons, snails and slugs have adapted it for creeping; the bivalves use it for burrowing; whilst in the squid and octopus it is adapted for swimming. All molluscs have gills, which in the bivalves are also used for filter feeding. Many molluscs have a calcareous shell, which often has a characteristic shape, pattern and coloration for each species. It is secreted by a special fold of tissue called the mantle, which is also involved in protecting the gills.

The feeding habits of molluscs are diverse. The bivalves pass water currents over their gills, from which they filter out minute particles of food; all the other molluscs use a feeding organ known as a radula. The grazing molluscs (such as chitons, slugs and some snails) have a long, flexible, file-like radula bearing many small teeth, with which they scrape the surface of rocks and seaweeds. With carnivorous molluscs (including some snails, squid and octopus) the radula is reduced to a few, much larger teeth.

Of the seven classes of mollusc, four are included here.

The chitons (class Polyplacopora) are all marine. Their shells are made up of eight dorsal plates (valves), which overlap from front to back and are embedded in a muscular girdle at the sides. Underneath there is a single, large foot with a mouth at the front and an anus at the back. The gills are confined to a groove between the foot and the girdle. Chitons attach themselves to rocks and are often identified by the sculpturing on the shell and girdle.

Gastropods (class Gastropoda – which includes slugs and snails) all move by the use of a single, often large, flattened foot. The snails have a single shell, which is usually coiled, or simply conical, as with the limpets. The type and extent of coiling, the sculpturing and coloration of the shell are all used for identification. The slugs have no shell, but are often highly coloured. The coloration is somewhat variable and so can only be used for identification together with the distribution and number of tentacles and gills.

The bivalves (class Bivalvia), which includes the mussels and oysters, all have a shell comprising two hinged valves. They are filter feeding animals and many are burrowers, using a modified, plough-shaped foot to dig into sand or mud. Other species attach themselves to stones, rocks or other permanent structures; mussels, for example, use their foot to produce sticky anchoring threads. A few species, such as the queen scallop, are free-living and able to swim by 'jet propulsion', by opening and then forcibly closing their valves. Shell shape, sculpturing and hinge detail, together with habitat, are important in identification.

Cuttlefish, squid and octopus are all members of the class Cephalopoda. They are only found on the shore when stranded by the tide or trapped in rockpools. They are highly active swimmers with well developed eyes, which they use for hunting.

Leptochiton asellus

(*North Sea, English Channel, Atlantic.*) To 2cm in length. Oval in shape with broad shell valves and a narrow girdle. 8-13 gills extend only half way up groove. Valves fragile, smooth, girdle granular, covered by relatively large overlapping plates. Beige to straw coloured, maybe with brown flecks, or black. Lower shore.

Ischnochiton albus

(*North Sea, Baltic Sea, English Channel, Atlantic.*) To 1.5cm in length. Elongate oval, almost cigar shaped with markedly keeled shell valves and a narrow girdle. 10-16 gills extend half way up groove. Valves smooth and shiny, girdle coarsely granular. White to off-white or maybe black. Lower shore.

Lepidochitona cinereus

(*North Sea, Baltic Sea, English Channel, Atlantic.*) To 2.8cm in length. Moderately keeled shell valves and a narrow girdle. 15-22 gills extend entire length of groove. Valves with granular appearance. Girdle finely granular and fringed by spines. Colour variable, including white, black, green, pink, blue and orange. Girdle often banded. Middle shore downwards.

Callochiton achatinus

(*North Sea, Baltic Sea, English Channel, Atlantic.*) To 3cm in length. Flattened shell valves, slightly curved with slight keel and broad girdle. 20-25 gills extend entire length of groove. Valves slightly granular and with hand lens can see regularly arranged tiny black dots. Girdle with snake-skin appearance. Reddish-brown but maybe greenish or pink, girdle brick red, often yellowish bands at four corners. Lower shore.

Tonicella marmorea

(*North Sea, Atlantic.*) To 4.5cm in length. Oval in shape, shell valves flattened but keeled, girdle moderately narrow. 20-28 gills usually extend more than half way up groove. Valves almost smooth, girdle leathery with fringe of spines. Valves, various shades of red with brown and cream mottling, girdle banded variously red, green and orange. Lower shore.

Tonicella rubra

(*North Sea, English Channel, Atlantic.*) To 2cm in length. Oval in shape, shell valves round and moderately keeled, girdle narrow. 10-16 gills extend half way up groove. Valves smooth, girdle finely granular in appearance, fringed by line of golden spines. Shades of pink, often with brick-red patches, girdle banded cream and orange-pink. Lower shore.

Acanthochitona crinitus

(*North Sea, English Channel, Atlantic.*) To 3.5cm in length. Elongate oval in shape with narrow, keeled valves, broad girdle. 10-15 gills extend half way up groove. Valves with distinctive sculpturing. Girdle with 18 tufts of bristles. Colour highly variable, girdle often greenish-yellow. Upper shore downwards.

Acanthochitona fascicularis

(*North Sea, English Channel, Atlantic.*) To 6cm in length. Oblong in shape, valves narrow and keeled, girdle broad and fleshy. 14-20 gills extend half way up groove. Valves with narrow, smooth central region, lateral regions with densely packed, round papillae. Girdle with 18 tufts of short bristles. Usually olive-green with reddish patches. Lower shore.

Haliotis tuberculata

Green Ormer

(*English Channel, Atlantic.*) To
10cm in length. Flattened shell with
spiral appearance. Series of
openings on surface, usually five.
Green, brown or red, with mother-
of-pearl inside. Lower shore.

Diodora graeca

Keyhole Limpet

(*North Sea, English Channel,
Atlantic.*) To 4cm in length.
Conical shell with no sign of
spiralling. Fine cross-ribbed
sculpturing and a hole in apex of
shell. Animal often extends out
around margin of shell. Greyish or
beige. Lower shore.

Eumarginula reticulata

Slit Limpet

(*North Sea, English Channel,
Atlantic.*) To 2cm in length. Shell,
conical with apex slightly curved
over, ribs running from apex to
margin and with distinctive notch
in front margin. Whitish-green or
yellow. Lower shore.

Tectura testudinalis

Tortoiseshell Limpet

(*North Sea, Baltic Sea, Atlantic.*) To
2.5cm in length. Shell, flattened
cone with smooth surface.
Tortoiseshell appearance with
delicate red and brown markings.
Rim of large foot greenish. Lower
shore.

Tectura virginica

White Tortoiseshell Limpet

(*North Sea, Baltic Sea, English Channel, Atlantic.*) To 1.5cm in length. Shell, flattened smooth cone with apex towards the front and relatively solid appearance. White-pinkish tortoiseshell markings. Lower shore.

Patella depressa

(*English Channel, Atlantic.*) To 4.5cm in length. Shell, flattened oval cone, finely ribbed. Greyish-beige in colour, though margin may be marked with contrasting dark rays. Internal scar left by animal creamy-orange. Middle shore.

Patella ulyssiponensis

(*North Sea, English Channel, Atlantic.*) To 7cm in length. Shell, bluntly pear-shaped, flattened cone with apex near to front, only finely ribbed. No contrasting marginal rays. Shell greyish-beige, foot orange, internal scar white. Prefers sheltered sites. Upper shore in rockpools, otherwise middle shore downwards.

Patella vulgata

Common Limpet

(*North Sea, English Channel, Atlantic.*) To 8cm in length. Shell, oval cone, typically tall, strong, irregular ribbing. Shell greyish-beige, often with barnacles growing on it, foot olive-green, internal surface yellowish with brownish to white scar. Upper shore downwards.

Helcion pellucida
Blue-rayed Limpet

(*North Sea, English Channel, Atlantic.*) To 2cm in length. Shell, oval, smooth and semi-transparent with slightly hooked, off-centre apex. Shell orange with rays of brilliant blue dots, colours fade in older animals. On *Laminaria spp*. Lower shore.

Monodonta lineata
Thick Topshell

(*English Channel, Atlantic.*) To 2.5cm in height and breadth. Shell, round cone with blunt six-whorl spiral. Aperture lined with mother-of-pearl and bearing single tooth. Greyish-green with pattern of purple zigzag streaks. Middle shore.

Gibbula magus

(*North Sea, English Channel, Atlantic.*) To 3cm in breadth. Shell, flattened cone with up to eight whorls and conspicuous umbilicalis. Shell surface roughly sculptured. Pale yellow with red or purple markings. Lower shore.

Gibbula cineraria
Grey Topshell

(*North Sea, English Channel, Atlantic.*) To 1.5cm in height. Shell, conical with varying degree of flattening, up to seven whorls and small umbilicalis. Light greyish-pink with darker stripes. Middle shore downwards.

Calliostoma zizyphinum

Common or Painted Topshell

(*North Sea, English Channel, Atlantic.*) To 2.5cm in height and breadth. Shell, pyramidal cone with flattened base, pointed spire, glossy surface, up to nine whorls, no umbilicalis. Pale straw colour with red streaks, or white. Lower shore.

Gibbula umbilicalis

Purple Topshell

(*North Sea, English Channel, Atlantic.*) To 1.5cm in breadth. Shell, flattened cone, up to seven whorls and conspicuous umbilicalis. Light brownish-green with red or purple rays. Middle shore.

Lacuna vincta

Banded Chink Shell

(*North Sea, Baltic Sea, English Channel, Atlantic.*) To 1cm in height. Shell, conical, pointed apex, thin, with transparent glossy appearance. Up to six whorls and deep umbilicalis. White to yellow with red spiral bands. Lower shore.

Littorina littorea

Edible Periwinkle

(*North Sea, Baltic Sea, English Channel, Atlantic.*) To 3cm in height. Shell, moderately tall cone, thick, with fine spiral ridges. Up to eight whorls. Greyish-bluish-black or red. Upper/middle shore downwards.

Littorina neritoides

Small Periwinkle

(*North Sea, English Channel, Atlantic.*) To 7mm in height. Shell, pointed cone, smooth and fragile. Blue-black. Found on more exposed shores. Upper shore and splash-zone.

Littorina saxatilis

Rough Periwinkle

(*North Sea, Baltic Sea, English Channel, Atlantic.*) To 1.5cm in height. Shell, sharply pointed cone, up to eight whorls, with sculptured surface. Colour variable, brown, red, grey, green, beige, maybe banded. Upper to middle shore.

Turritella communis

Tower Shell

(*North Sea, English Channel, Atlantic.*) To 6cm in height. Shell, long tapering cone, pointed apex, small aperture. Up to 19 whorls, sculptured with spiral ridges. Colour variable, red, brown, yellow, white. Lower shore.

Littorina obtusata

Flat Periwinkle

(*North Sea, Baltic Sea, English Channel, Atlantic.*) To 1.5cm in height. Shell, flattened, blunt spiral, thick, with very fine surface sculpturing and exceptionally large aperture. Colour variable, yellow, green, brown, red, orange, maybe banded. Middle shore downwards.

Hydrobia ulvae

Laver Spire Shell

(*North Sea, Baltic Sea, English Channel, Atlantic.*) To 1cm in height. Shell, elongate cone with blunt apex, thick and smooth with up to six whorls. Reddish-brown, yellowish-green. Middle shore.

Bittium reticulatum

Needle Shell

(*North Sea, Baltic Sea, English Channel, Atlantic.*) To 1.5cm in height. Shell, tapering cone, pointed apex, no umbilicalis. Up to 16 whorls, nodular surface. Reddish-brown, nodules maybe white. Lower shore.

Trivia monacha
European Cowrie

(*North Sea, English Channel, Atlantic.*) To 1.2cm along slit. Shell, oval, spire hidden. Sculptured with 20–25 fine ribs. Pinkish-brown with brown dots above, paler below. Lower shore.

Lunatia poliana
Common Necklace Shell

(*North Sea, English Channel, Atlantic.*) To 1.8cm in height. Low spire, up to six whorls, last one by far the largest. Large, thick-edged aperture. Umbilicalis half closed. Buff with reddish or brownish streaks. Lower shore.

Crepidula fornicata
Slipper Limpet

(*North Sea, English Channel, Atlantic.*) To 4cm in length. Shell, solid, oval, apex spirally curved to right projecting beyond back end of rim. Change sex as grow older, young male, old female. Grow stacked on top of each other, oldest at bottom. Brownish-yellow with reddish markings. Lower shore.

Velutina velutina
Velvet Shell

(*North Sea, English Channel, Atlantic.*) To 2cm in height. Shell, oval, semi-transparent, covered in velvety layer with up to three and a half whorls. Pale brown. Often found on sea squirts. Lower shore.

Nucella lapillus
Dogwhelk

(*North Sea, English Channel, Atlantic.*) To 4cm in height. Shell with flattened spiral ridges, up to five whorls. Aperture with siphon groove. Colour variable, white, yellow, purple, brown, maybe banded. Middle shore downwards.

Ocenebra erinacea

Sting Winkle

(*North Sea, English Channel, Atlantic.*) To 5cm in height. Shell, massive with rough sculpturing of longitudinal and spiral ribs. Up to ten whorls, siphon groove present in young. Yellowish-white with red and brown streaks. Lower shore.

Buccinum undatum

Common Whelk

(*North Sea, Baltic Sea, English Channel, Atlantic.*) To 8cm in height. Shell, conical with pointed apex, opaque, with five spiral and five longitudinal ribs. Up to eight whorls, large aperture with short siphon groove. Yellowish-white, reddish or purple. Lower shore.

Hinia reticulata

Netted Dogwhelk

(*North Sea, Baltic Sea, English Channel, Atlantic.*) To 3.5cm in height. Shell, conical with strong longitudinal ribs and fine spiral ribs creating tubercles. Up to ten whorls, aperture with toothed outer lip. Brown. Lower shore.

Acteon tornatilis

(*North Sea, English Channel, Atlantic.*) To 2cm in height. Shell, oval, spire to pointed apex, up to seven whorls, the last one by far the largest. Pale pink to brown with white banding. Lower shore.

Philine aperta

(*North Sea, Baltic Sea, English Channel, Atlantic.*) To 2cm in length. Body divided into four tubes, enclosing reduced, semi-transparent, glossy shell. Squarish-oval in shape. Translucent white. Lower shore.

Leucophytia bidentata

(*North Sea, English Channel, Atlantic.*) To 1cm in height. Shell conical, pointed, with up to seven whorls and two projections on inner wall of aperture. Opaque white. Upper shore.

Otina ovata

(*English Channel, Atlantic.*) To 5mm in height. Shell rather thin, semi-transparent, two whorls, the second forming most of shell. Reddish brown or purple. Often found in dead barnacle shells and cracks. Upper shore.

Facelina coronata

(*English Channel, Atlantic.*) To 2.5cm in length. No shell. Body slender, tapering with two pairs of dissimilar tentacles at head end and six rows of appendages across back. Whitish, appendages rose-red with pale blue tips. Lower shore.

Aeolidia papillosa
Common Grey Sea-slug

(*North Sea, English Channel, Atlantic.*) To 9cm in length. No shell. Body broadly ovate with two pairs of tentacles. Dense rows of appendages cover back with naked line down centre. Greyish-brown, appendages brown. Upper shore downwards.

Elysia viridis

(*North Sea, Baltic Sea, English Channel, Atlantic.*) To 3cm in length. No shell. Body, elongate-oval with lateral lobes, two tentacles and no gills. Green with white spots. Found on green seaweeds, particularly *Codium tomentosum* (p.44). Middle shore downwards.

Cadlina laevis

(*North Sea, Atlantic.*) To 2.5cm in length. No shell. Body, rounded oval in shape with tubercles over surface, two head tentacles, ring of five tripinnate gills at rear. Opaque white with row of yellow dots down each side. Lower shore.

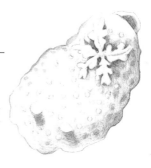

Archidoris pseudoargus

Sea Lemon

(*North Sea, English Channel, Atlantic.*) To 7cm in length. No shell. Body, elliptical with tubercles over surface, two head tentacles, ring of nine tripinnate gills towards rear. Yellow with green, pink or brown blotches. Lower shore.

Jorunna tomentosa

(*North Sea, English Channel, Atlantic.*) To 5cm in length. No shell. Body, ovate and slightly flattened with small tubercles over surface, two head tentacles, ring of 15 tripinnate gills towards rear. Yellowish with brown spots, gills somewhat paler. Lower shore.

Musculus discors

Green Crenella

(*North Sea, English Channel, Atlantic.*) To 1.2cm in length. Shell, oval, compressed, thin with longitudinal ribs. Valves similar with two muscle scars. Greenish, dull. Middle shore downwards.

Anomia ephippium

Common Saddle Oyster

(*North Sea, English Channel, Atlantic.*) To 6cm in length. Shell, flat, crescent-shaped right (lower) valve. Umbo near margin, three muscle scars on inside of upper shell. Dull white outside, glossy inside. Lower shore.

Mytilus edulis

Common Mussel

(*North Sea, Baltic Sea, English Channel, Atlantic.*) To 10cm in length. Shell, roughly triangular, two muscle scars, front one large, back one small. Glossy when small. Dark blue or brownish-yellow. Attached to rocks or shells by threads. Upper shore downwards.

Modiolus modiolus

Horse Mussel

(*North Sea, English Channel, Atlantic.*) Sometimes to 15cm in length. Shell, thick, valves similar, two muscle scars, front one small, back one large. Purplish-brown. Often associated with *Laminaria spp.* Lower shore.

Crassostrea gigas

Portuguese Oyster

(*North Sea, English Channel, Atlantic.*) To 7cm in length. Shell, heavily sculptured, valves dissimilar, deep left (lower) valve, flat right (upper) valve. Brownish to olive-green, internally whitish. Bottom of lower shore.

Ostrea edulis

Common European Oyster

(*North Sea, English Channel, Atlantic.*) To 7.5cm in length. Shell, roughly round, margin crenulated, valves dissimilar, right (upper) one flatter, one muscle scar. Often bored by sponge, *Cliona celata* (p.59). Yellowish brown. Bottom of lower shore.

Chlamys varia

Variegated Scallop

(*North Sea, English Channel, Atlantic.*) To 4.5cm in length. Shell, almost oval, valves similar with 25-30 ribs bearing small teeth, ears unequal, 30 black dots (ocelli) round edge of mantle. Either free-living or attached by threads. Colour variable, brown, yellow, white, red, purple. Lower shore.

Aequipecten opercularis

Queen Scallop

(*North Sea, English Channel, Atlantic.*) To 8cm in length. Shell, circular, thin, valves dissimilar, right (lower) valve flatter, 20 rounded ribs, ears only slightly unequal. 35-40 ocelli (white dots surrounded by black rings) round edge of mantle. Active swimmer. Colour variable, brown, yellow, orange, red, purple. Lower shore.

Pecten maximus

Great Scallop

(*North Sea, English Channel, Atlantic.*) To 15cm in length. Shell nearly circular, valves dissimilar, right (lower) valve convex, left (upper) valve flat. Rounded ribs and concentric growth lines, ears equal. Yellow or reddish-brown. Often free-swimming. Lower shore.

Parvicardium exiguum

Little Cockle

(*North Sea, English Channel,
Atlantic.*) To 1.2cm in length. Shell
roughly triangular, valves similar,
20 flattened ribs and crenulated
margin. Shell, dull, brownish
outside, greenish-white inside.
Lower shore.

Cerastoderma edule

Common Cockle

   \

(*North Sea, English Channel,
Atlantic.*) To 5.5cm in length. Shell,
rounded triangle, valves similar, 24–
28 ribs and many fine concentric
lines, crenulated margin. Obvious
muscle scars internally. Shell off-
white to yellowish-brown outside,
white with brown markings inside.
Lower shore.

Mactra staltorum

Rayed Trough Shell

(*North Sea, English Channel,
Atlantic.*) To 5cm in length. Shell,
long, convex, quite light, glossy,
valves similar, hinge complex and
producing sharp beak. Two
obvious, almost equal, muscle scars.
Yellowish-white with concentric
reddish-brown lines, internally
purplish-white. Lower shore.

Lutraria lutraria

Common Otter Shell

(*North Sea, English Channel,
Atlantic.*) To 14cm in length. Shell,
elliptical, valves equal, hinge
complex, irregularly striated,
concentric growth lines, smooth
margin. Two deep muscle scars
evident. Glossy yellowish-white,
periostracum greenish-brown
(when present). Lower shore.

Ensis ensis

(*North Sea, English Channel, Atlantic.*) To 12cm in length. Shell, curved, valves similar, tapering slightly towards back end. Glossy, yellowish-white with brown or reddish streaks. Lower shore.

Ensis siliqua

Pod Razor Shell

(*North Sea, English Channel, Atlantic.*) To 20cm in length. Shell with parallel sides, not tapering, valves similar. Glossy, whitish, finely lined with reddish markings, yellowish-green periostracum. Lower shore.

Angulus tenuis

Thin Tellin

(*North Sea, Baltic Sea, English Channel, Atlantic.*) To 2.5cm in length. Shell, flattened, thin, glossy, valves almost equal, often with fine concentric lines. Beautifully coloured, white, yellow, orange or red. Middle shore downwards.

Fabulina fabula

(*North Sea, Baltic Sea, English Channel, Atlantic.*) To 2cm in length. Shell, flattened, thin valves similar, angular and pointed at back end. Fine concentric lines, evident under hand lens, on right valve. White, yellow or orange. Lower shore.

99

Macoma balthica

Baltic Tellin

(*North Sea, Baltic Sea, English Channel, Atlantic.*) To 2.5cm in length. Shell, globular with close-set, concentric lines, valves similar. Opaque, pink, purple, brown, white. Middle shore downwards.

Donax vittatus

Banded Wedge Shell

(*North Sea, English Channel, Atlantic.*) To 3cm in length. Shell, compressed, opaque and polished with fine concentric lines, valves similar, toothed edge. Yellow, brown or purple, inside violet. On exposed shores. Lower shore.

Gari depressa

Large Sunset Shell

(*North Sea, English Channel, Atlantic.*) To 4cm in length. Shell, oval, compressed, rather solid, opaque, glossy with fine concentric ridges. Valves similar, with gap at back end. Yellowish-white with purple-brown rays. Lower shore.

Scrobicularia plana

Peppery Furrow Shell

(*North Sea, Baltic Sea, English Channel, Atlantic.*) To 6cm in length. Shell, compressed, opaque, fragile. Valves similar. Greyish-white outside, white inside. Middle shore downwards.

Paplica rhomboides

Banded Carpet Shell

(*North Sea, English Channel, Atlantic.*) To 5cm in length. Shell, oval, glossy, flattened with irregular concentric ribs, valves similar. Yellowish-white with red-brown zigzag markings. Bottom of lower shore.

Chamelea gallina
Striped Venus

(*North Sea, English Channel, Atlantic.*) To 4cm in length. Shell, triangular, valves similar, concentric ribs crossed by fine longitudinal lines. Yellowish-white, three reddish-brown rays, chalky white inside. Lower shore.

Mya arenaria
Sand Gaper

(*North Sea, Baltic Sea, English Channel, Atlantic.*) To 15cm in length. Shell, oblong, right valve more convex than left, concentric growth lines. Ashy-grey to brown, inside chalky white. Lower shore.

Tapes decussatus
Cross-cut Carpet Shell

(*North Sea, English Channel, Atlantic.*) To 4.5cm in length. Shell, squarish, massive, valves similar, tubercles on surface. Yellowish with purple markings. Lower shore.

Mya truncata
Blunt Gaper

(*North Sea, English Channel, Atlantic.*) To 6.5cm in length. Shell, solid, right valve more convex than left, concentric growth lines. Gape at back end. Greyish-white, inside chalky white. Middle shore downwards.

Pholas dactylus
Common Piddock

(*English Channel, Atlantic.*) To 13cm in length. Shell, elongate, delicate, valves similar, many rows of points formed where concentric and longitudinal rays overlap. Wide ventral gape at front. White-grey, inside white. Found in peat, wood and hard sand. Lower shore.

Sepia officinalis
Common Cuttlefish

(*North Sea, English Channel, Atlantic.*) To 30cm in length. Often in shallow water, over *Zostera spp* (p.130). Internal cuttle bone often washed ashore.

Loligo forbesi
Common Squid

(*North Sea, Baltic Sea, English Channel, Atlantic.*) To 60cm in length. Pen-like internal shell. Very active, fast swimmer. Only found on shore when stranded.

Eledone cirrhosa
Lesser Octopus

(*North Sea, English Channel, Atlantic.*) To 50cm in length. Body, bag-shaped, internal shell reduced to pair of minute stylets. Reddish-brown above, white below. Bottom of lower shore.

Sea Mats (Bryozoa)

These small, sessile (stationary), colonial animals are rather inconspicuous and typically either off-white, beige or a shade of orange. They are common on rocky shores, attached to boulders and seaweeds (in particular *Fucus serratus* and the various Laminarians).

Sea mats are filter feeders, using a ring of tentacles to trap the minute particles of food suspended in the water. There are many species of sea mat, but in order to identify them you would need to use a microscope to look at various external features, including the position of the opening, whether or not the opening has a lid and the presence or absence of bristles. For this reason, no attempt is made here at species identification. Instead, some examples of the different growth forms are given below which will enable a specimen to be identified as a sea mat.

Growth Form A: encrusting patches, irregular in shape and size. Maybe rough to the touch, with honeycomb appearance. Middle shore downwards.
Growth Form B: tufted, much branched, maybe several centimetres in length, attached at the base to either rock or seaweed.
Growth Form C: much branched (often dichotomously), flattened fronds with rounded tips and a basal attachment point. Zooids regularly arranged on both sides of frond and only slightly rough to the touch. Usually found as detached specimens, washed ashore.

Feather Stars, Starfish, Brittle Stars, Sea Urchins and Sea Cucumbers (Echinodermata)

The echinoderms are essentially radially symmetrical, marine animals with a central mouth, normally on the underside, and the anus on the opposite side. Most have five rays (or multiples of five) and use tube feet for locomotion. These feet work hydrostatically and usually occur in double rows along each ray. Echinoderms have a skeleton composed of calcareous plates and spines, some of which may be fused together to form a shell, or test.

Starfish (class Asteroidea), brittle stars (class Ophiuroidea) and feather stars (class Crinoidea) all have the classic star shape, with central disc and radiating arms. Starfish are carnivores and hunt by smell, while brittle stars are scavengers, using the same opening for mouth and anus. In contrast, the feather stars use their tube feet to filter feed, and have both the mouth and anus on the upper side of the body. With all of these animals, when attacked, one or more of the radiating arms may be sacrificed and subsequently regrown.

Sea urchins (class Echinoidea) are globular in shape – as if the rays had been curled up to form a ball – with well developed tests. They are grazers of seaweeds and encrusting animals. Their tube feet are used mainly for attachment; the spines are used for locomotion.

Sea cucumbers (class Holothuroidea) are cucumber shaped, with the front end often identified by the presence of modified tube feet, used to sweep food into their mouths.

Antedon bifida

Feather Star

(*North Sea, English Channel, Atlantic.*) To 15cm in diameter with arms spread. Concave central disc with 25 finger-like projections (cirri) which provide temporary attachment. Five pairs of feathery arms which wave in water. Red, pink, purple, orange, or yellow. Lower shore.

Asterina gibbosa

Cushion Star, Starlet

(*North Sea, English Channel, Atlantic.*) To 6cm in diameter. Five stubby arms radiate from slightly swollen disc. Stiff, rough texture. Greenish with tinge of yellow or red. Tips of arms curl up when disturbed. Lower shore.

Solaster endeca

Purple Sunstar

(*North Sea, Baltic Sea, Atlantic.*) To
40cm in diameter, more usually to
25cm. Normally nine or ten arms.
Hard and rough to touch with
close-set short spines. Violet or
yellowish-red. Lower shore.

Henricia sanguinolenta

(*North Sea, Baltic Sea, English
Channel, Atlantic.*) To 20cm in
diameter. Relatively small disc
produces five slender, almost
round, tapering, stiff arms. Usually
red, but may be purple or orange
on top, pale sandy colour
underneath. Lower shore.

Asterias rubens

Common Starfish

(*North Sea, Baltic Sea, English
Channel, Atlantic.*) To 50cm in
diameter (in sea), but usually to
15cm (on shore). Disc produces
typically five, plump, rounded,
bluntly tapering arms. Skin rather
flabby and rough with distinct line
of spines down each ray. Reddish-
brown or pale red or yellow.
Lower shore.

Marth. sterias glacialis

Spiny Starfish

(*North Sea, English Channel,
Atlantic.*) To 30cm in diameter,
occasionally to 80cm. Small disc
produces five gradually tapering
rays. Soft body with rough texture
due to covering of spines and
knobs. Greyish-green, yellowish or
reddish. Lower shore.

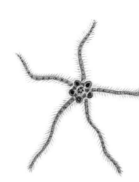

Ophiothrix fragilis

Common Brittle Star

(*North Sea, English Channel, Atlantic*.) To 10cm in diameter. Disc, very small with five rays of spines and two naked plates between rays. Arms, tapering, fragile (often broken), covered with scales, each joint with seven spines each side. Variable colour, violet, purple, red, yellow, grey. Lower shore.

Ophiocomina nigra

(*North Sea, English Channel, Atlantic*.) To 12.5cm in diameter. Disc, very small (to 2.5cm in diameter) and feels smooth. Rays relatively thin and tapering with five to seven long, erect spines at joints. Black, brown, maybe even pink or grey. Lower shore.

Amphiura brachiata

(*North Sea, English Channel, Atlantic*.) To 18cm in diameter. Disc small (to 1.2cm in diameter). Arms, very long and flexible, with eight to ten spines each side of joints near disc. Bluish or greyish-brown. Often only tips of arms showing above sand. Lower shore.

Ophiura ophiura

(*North Sea, Baltic Sea, English Channel, Atlantic*.) To 14cm in diameter. Disc, to 3.5cm in diameter, with distinct pairs of scales at origin of rays. Rays, tapering, three spines at joints, lying along ray. Ventral ray plates have pore-shaped groove between them. Red or reddish-brown, may be spotted. Lower shore.

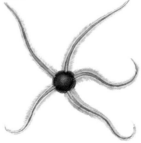

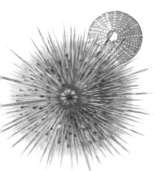

Psammechinus miliaris
Green Sea Urchin

(*North Sea, Baltic Sea, English Channel, Atlantic.*) To 5cm in diameter. Test covered in short (1.5cm), strong spines. Pale green with violet tips to spines. May have pieces of seaweed attached to spines. Middle shore downwards.

Echinus esculentus
Edible Sea Urchin

(*North Sea, English Channel, Atlantic.*) To 12cm in diameter, sometimes larger. Test almost spherical but flattened. Spines short (1.5cm) and blunt. Test, red with white spots at spine attachments, spines white or red with purple tips. Lower shore.

Paracentrotus lividus

(*Western English Channel, Atlantic.*) To 5cm in diameter. Test markedly flattened, numerous long (3cm), tapering, solid spines. Test dark green or brown, spines violet or brownish. May bore into rock and often covered by pieces of seaweed and shell. Middle shore downwards.

Strongylocentrotus droebachiensis

(*North Sea, North Atlantic.*) To 7.5cm in diameter. Test, low, but not flattened, spines short (2cm), striated with a rounded end. Test green or brown, spines green, red or violet. May bore into rock. Lower shore.

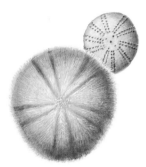

Echinocyamus pusillus
Pea Urchin

(*North Sea, Baltic Sea, English Channel, Atlantic.*) To 1.5cm in length. Test, oval, slightly depressed, covered in dense, short spines. Grey or green (dirty white when dead). Lower shore.

Echinocardium cordatum
Sea Potato

(*North Sea, English Channel, Atlantic.*) To 9cm in length. Test bears dense covering of soft, brittle spines, mainly short but some longer. Spines point backwards and are easily rubbed off. Straw coloured. Near-white and spineless when dead. Lower shore.

Holothuria forskali
Cotton Spinner

(*English Channel, Atlantic.*) To 20cm in length. Body covered by a thick, coarse skin. At head end, 20 much branched, retractible, yellow tentacles. Dorsally black or dark brown, ventrally yellowish, with three distinct rows of suckers. Lower shore.

Aslia lefevrie

(*English Channel, Atlantic.*) To 15cm in length. Body covered by thick, tough, leathery, wrinkled skin. At head end, ten feathery tentacles. Suckers on ventral surface arranged in double rows. Brownish-white to black. Lower shore.

Sea Squirts (Tunicata)

Unlikely as it may seem, sea squirts are near to vertebrates in evolutionary terms; the free-swimming larval stage possesses a notochord (backbone), just like vertebrates, although there is no sign of it in the adult.

The bottom-dwelling adult lives attached to rock, seaweed or other animals. The body is contained within an outer skin, or tunic, which is often very tough and almost transparent. There are two openings, or siphons, usually at the opposite end of the tunic from the point of attachment; the inhalant siphon is situated centrally, the exhalant siphon is offset to one side.

Water is drawn in the central siphon and passed over an extensive gill system, which absorbs oxygen and filters out food particles. Waste material is passed out of the exhalant siphon. Sea squirts may be solitary or colonial. When colonial, a number of individuals share a single exhalant siphon.

Clavelina lepadiformis
Light-bulb Tunicate

(*North Sea, English Channel, Atlantic.*) To 2cm in height. Stalked, occuring in clusters from a common base. Test, clear and gelatinous, inhalant and exhalant siphons close together. Yellow or brown pigment stripes show through test. Lower shore.

Didemnum maculosum

(*North Sea, English Channel, Atlantic.*) To 2mm thick. Colonial, forming a thick, rather rough, leathery film, may feel gritty to touch. Between five and nine individuals share a single exhalant siphon. Bluish-grey or violet with dark purple streaks. Lower shore.

Ciona intestinalis

(*North Sea, Baltic Sea, English Channel, Atlantic.*) To 12cm in height, 4cm in diameter. Solitary, siphons are unequal and close together. Test, attached at the bottom, more or less transparent. Contracts when touched. Test, greenish-grey with five obvious muscle bands, siphon tips yellow. Lower shore.

Ascidia mentula

(*North Sea, Baltic Sea, English Channel, Atlantic.*) To 10cm in height. Solitary. Test, flattened pear shape with rough, tough texture. Inhalant siphon at end. Milky-green colour with reddish tinge around siphons. Lower shore.

Distomus variolosus

(*North Sea, English Channel, Atlantic.*) To 1cm in height. Colonial. Test, rounded with rough appearance. Siphons withdrawn when tide out. Brick red to brown. Bottom of lower shore.

Botrylloides leachii

(*North Sea, English Channel, Atlantic.*) Colonial, of varying size. Encrusting, thick and fleshy, arranged in irregular rows. Colour variable, orange, yellow, bluish and grey. Middle shore downwards.

Ascidiella aspera

(*North Sea, Baltic Sea, English Channel, Atlantic.*) To 12cm in height. Solitary. Test, long and oval, rough appearance. Reddish-brown or grey. Lower shore.

Dendrodoa grossularia

(*North Sea, Baltic Sea, English Channel, Atlantic.*) To 2.5cm in height. Solitary or in groups. Test, squat, wrinkled with inhalant siphon at the top. Rusty, red or yellow. Lower shore.

Botryllus schlosseri

Star Sea Squirt

(*North Sea, English Channel, Atlantic.*) Colonial, of varying size. Encrusting, thick and fleshy. Colour variable, violet, green, yellow and red. Middle shore downwards.

Fish (Pisces)

Fish can be divided into two main groups, as defined by their skeleton; cartilaginous fish (which includes the sharks and rays) and bony fish (which include all the species normally found on the shore).

In common with many other vertebrates, the bony fish have two pairs of appendages; pectoral fins (which correspond to forelimbs) and pelvic fins (which correspond to hindlimbs). In addition, they have a number of unpaired fins. The head bears a well developed mouth, with a true jaw and teeth, a pair of large eyes and gills, covered by an opercular flap.

The fish that are commonly found on the shore show a number of adaptations to the intertidal environment. Many are shaped so as to make burrowing and squirming between stones or into crevices easier. Others have suckers for clinging onto rocks. Beware of relying on colour for identification; fish often change colour when alive and after death.

Lepadogaster lepadogaster

Cling Fish, Cornish Sucker

(*English Channel, Atlantic.*) To 7cm in length. Spatulate snout, thick body tapering to tail. Pectoral fins rounded, pelvic fins modified to form sucker, dorsal fin longer than anal fin, both merge with tail. Colour variable, red or purplish, lighter on ventral surface, two large oval spots, black surrounded by white, on head. Middle shore downwards.

Lophius piscatorus

Anglerfish

(*North Sea, English Channel, Atlantic.*) To 2m in length. Mouth enormous with jutting lower jaw, body tapering to tail. Eyes on top of head pointing upwards, with three long spines in row between eyes. Small dorsal and anal fins. Brownish, lighter beneath. Grotesque. Lower shore.

Ciliata mustela

Five-bearded Rockling

(*North Sea, Baltic Sea, English Channel, Atlantic.*) To 20cm in length. Four barbels above upper jaw, one below lower jaw. Body, tapering, first dorsal fin reduced except for leading ray, second dorsal fin long. Light brown, paler underneath, fins darker, orange near base. Lower shore.

Gaidropsarus mediterraneus

Shore Rockling

(*English Channel, Atlantic.*) To 25cm in length. Two barbels above upper jaw, one below lower jaw.

Body tapering to tail, first dorsal fin reduced except for leading ray, second dorsal fin long, anal fin shorter. Reddish-grey with irregular dark spots, white below. Lower shore.

Gasterosteus aculeatus

Three-spined Stickleback

(*North Sea, Baltic Sea, English Channel, Atlantic.*) To 7cm in length. Head pointed, body plump, narrowing abruptly to stalk for tail. Three dorsal spines. Dorsal fin slightly longer than anal fin. Pelvic fins very much reduced. Grey with green reflections and bronze sides. Lower shore.

Spinachia spinachia

Fifteen-spined Stickleback

(*North Sea, Baltic Sea, English Channel, Atlantic.*) To 15cm in length. Head with long snout, body tapering to long stalk and fan-shaped tail. Flanks with raised ridge of plates, 15 spines on back. Dorsal and anal fins equal, pelvic fins much reduced. Colour variable, olive-green on back, silver flanks, with dark brown raised plates. Lower shore.

Fish

Nerophis ophidion

Straight-nosed Pipefish

(*North Sea, Baltic Sea, English Channel, Atlantic.*) To 30cm in length. Snout about half length of head, with ridge running to eyes. Body very thin, short dorsal fin about half way along body, no pectoral or tail fins. Greenish with irregular pale markings. Lower shore.

Syngnathus acus

Greater Pipefish

(*North Sea, Baltic Sea, English Channel, Atlantic.*) To 45cm in length. Snout, tapering, more than half head length. Body tapers suddenly after dorsal fin. Dorsal fin, pectoral fins and small tail fin. About 20 body rings between pectoral and dorsal fins. Brownish, paler beneath. Lower shore.

Cyclopterus lumpus

Lump Sucker

(*North Sea, English Channel, Atlantic.*) To 50cm in length. Mouth relatively small, body deep and rounded. Overall rough appearance. Four rows of tubercles on each side. Front dorsal fin in form of a row of tubercles. Pelvic fins adapted to form sucker. Colour variable. Lower shore.

Liparis montagui

Montagu's Sea-snail

(*North Sea, Baltic Sea, English Channel, Atlantic.*) To 10cm in length. Head, blunt. Dorsal fin longer than anal fin, both distinct from tail fin. Pectoral fins extend ventrally, pelvic fins form sucker. No scales. Purplish or yellow-brown with dark spots. Middle shore downwards.

112

Crenilabrus melops

Corkwing Wrasse

  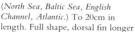

(North Sea, Baltic Sea, English Channel, Atlantic.) To 20cm in length. Full shape, dorsal fin longer than anal fin. Both dorsal and anal fins distinct from tail fin. Dark spot below lateral line just before tail. Colour and pattern highly variable, often three or four lines over face. Lower shore.

Lipophrys pholis

Shanny

(North Sea, English Channel, Atlantic.) To 12cm in length. Much flattened face, with distinct brow over eyes. Indistinct dark blotch at front of dorsal fin. Colour variable, olive-green or dark green with black markings. Lower shore.

Blennius ocellaris

Butterfly Blenny

(English Channel, Atlantic.) To 15cm in length. Heavy head. Body compressed sideways. Conspicuous tentacles above eyes. Greyish body with banding and characteristic black spot, ringed by white, on dorsal fin. Bottom of lower shore.

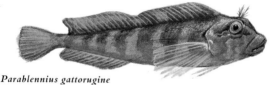

Parablennius gattorugine

Tompot Blenny

(English Channel, Atlantic.) To 20cm in length. Short, blunt head. Divided tentacle above each eye. Dorsal fin, long with undulating profile. Olive or brownish-grey, pale beneath with faint banding. Lower shore.

113

Zoarces viviparus

Eelpout, Viviparous Blenny

(*North Sea, Baltic Sea, English Channel, Atlantic.*) To 40cm in length. Body, long and tapering to

sharp point, no tail fin, dorsal and anal fins continuous. Notch in tail end of dorsal fin. Pelvic fins reduced. Olive with dark patches, pectoral fins orange. Lower shore.

Pholis gunnellus

Butterfish

(*North Sea, Baltic Sea, English Channel, Atlantic.*) To 20cm in length. Head, small and rounded, body flattened sideways. Dorsal fin

continuous for length of body, anal fin about half length of body, small tail fin. Pelvic fins very much reduced. Bronze colour with row of black dots half on dorsal fin, half on body. Lower shore.

Ammodytes tobianus

Lesser Sand-eel

(*North Sea, Baltic Sea, English Channel, Atlantic.*) To 20cm in length. Head, wedge shape with large mouth, body long and

slender. Dorsal fin long, anal fin less than half length of dorsal fin, distinct forked tail fin. Greenish-blue or greenish-yellow on back, silvery on flanks. Lower shore.

Pomatoschistus minutus

Sand Goby

(*North Sea, Baltic Sea, English Channel, Atlantic.*) To 8cm in length. Blunt, round head, gradually tapering body to moderately long tail stalk. Two

distinct dorsal fins, front one tallest at front end, back one similar to anal fin. Colour variable, sandy to grey with various blotches and distinctive dark spot at back end of front dorsal fin. Lower shore. (A number of closely related species.)

Mammals (Mammalia)

The only completely marine mammal is the whale, which never resorts to the land. Seals may spend weeks or months continuously at sea, but still return to land in order to breed, and at other times just to bask in the sunshine. They are much more at home in the water, having lost the ability to move with any agility on land. Otters are as at home on land as in the water. Those found on the seashores of Britain and Northern Europe are the same as those found in the rivers and lakes.

Halichoerus grypsis

Grey Seal

(*North Sea, Baltic Sea, English Channel, Atlantic.*) To 3m in length. Head with distinctive 'Roman' nose. Mottled, dark grey on fawn or pale grey on a darker background. Found on shore during autumn (for breeding) and spring (for moult).

Phoca vitulina

Common Seal

(*North Sea, Atlantic.*) To 1.9m in length. Small, fat, round-headed seal with a 'snub' nose. Leaden-grey or sandy with dark spots or dark grey with light spots. Pups born in summer. Tends to haul-out on sand banks exposed at low water rather than shores. Favours more sheltered sites than Grey Seal. Found in estuaries and sea lochs.

Lutra lutra

Otter

(*North Sea, English Channel, Atlantic.*) To 1.4m in length. Head with blunt muzzle. Long body with long tapering tail. Webbed feet. Dark brown with white throat and chest. Secretive animal, generally nocturnal but maybe active during day in more isolated areas.

Birds (Aves)

The seashore and the skies above are excellent for observing many different species of bird, including seabirds, waders and wildfowl. Like all the other seashore animals, they are dependent on the tidal cycle, whether it be for feeding, roosting or nesting. In addition, many of them have seasonal cycles of migration, so that they are only present on the shore at one particular time of year.

Calidris maritima

Purple Sandpiper

(*North Sea, Baltic Sea, English Channel, Atlantic.*) To 22cm in length. Bill, black with yellowish base, slightly curved. Short, dull yellow legs. Summer – rufous with white fringes above, blackish streaks and patches on breast and flanks. Winter – slate grey and blackish. White throat and belly. Rocky shores.

Calidris canutus

Knot

(*North Sea, English Channel, Atlantic.*) To 27cm in length. Bill, very short, black, straight. Greenish, short legs. Summer – head and underparts rufous, back blackish with chestnut fringes, wing coverts grey. Winter – mainly grey above, white below. Estuaries.

Calidris alpina

Dunlin

(*North Sea, Baltic Sea, English Channel, Atlantic.*) To 19cm in length. Bill, longish, black, curved at tip. Shortish black legs. Summer – black belly, rufous back. Winter – pale grey, streaked breast. Feeds in flocks. Saltmarshes, muddy and sandy shores.

Calidris alba

Sanderling

(*North Sea, English Channel, Atlantic.*) To 22cm in length. Bill, shortish, black, straight. Black legs. Summer – head and breast rufous, spotted black, belly white. Winter – white but for pale grey crown and back, leading wing edge black. Sandy and muddy shores.

Arenaria interpres

Turnstone

(*North Sea, Baltic Sea, English Channel, Atlantic.*) To 24cm in length. Bill, stubby, short, black. Legs orange. Summer – white head and underparts with black pattern on face and breast. Back, distinctive orange-chestnut with black-brown. Winter – head, underparts and breast, blackish-grey. All shore types.

Charadrius hiaticula

Ringed Plover

(*North Sea, Baltic Sea, English Channel, Atlantic.*) To 20cm in length. Bill, stubby, black tipped, orange base. Legs orange. Summer – black face mask and breast band, belly and face white, back brown. Winter – mask and breast band more brown. Shingle and sandy shores, mudflats in winter.

Vanellus vanellus

Lapwing, Peewit

(*North Sea, Baltic Sea, English Channel, Atlantic.*) To 32cm in length. Bill, stubby, black. Legs orange. Summer – face, throat and breast black, back and side of head and belly white, back bronzy green, undertail cinnamon, tail white with black bar. Dark head crest. Winter – same as summer plumage, but face less black and throat white. Coastal marshes, mudflats and farm land.

Haemotopus ostralegus

Oystercatcher

(*North Sea, Baltic Sea, English Channel, Atlantic.*) To 45cm in length. Bill, stout, long, orange. Legs pink. Summer – glossy black head, throat, breast and back. White underparts, tail white with black band. Orange-red eye. Winter – similar but duller and white throat band. Open shores, rocky, sandy and muddy.

Birds

Tringa totanus
Redshank

(*North Sea, Baltic Sea, English Channel, Atlantic.*) To 30cm in length. Bill, longish, straight, red base, blackish tip. Legs red. Summer – brown above with dark streaks and bars, white below, mottled and barred brown. Winter – uniform greyish above, breast less mottled and barred. Estuaries and saltmarshes.

Tringa erythropus
Spotted Redshank

(*English Channel, Atlantic.*) To 32cm in length. Bill, long, slender, black with reddish base. Legs, long, dark red. Summer – black, spotted white above. Winter – grey above, white below. Estuaries and saltmarshes.

Tringa hypoleucos
Common Sandpiper

(*North Sea, Baltic Sea, English Channel, Atlantic.*) To 20cm in length. Bill, shortish, pale base. Legs, short, grey-green. Summer – bronze-brown above with dark marks, olive chest, well streaked, belly white, white eyebrow. Winter – uniform olive-brown above, chest only slightly streaked. Estuaries and saltmarshes.

Tringa nebularia
Greenshank

(*North Sea, English Channel, Atlantic.*) To 32cm in length. Bill, long, stout, slightly upturned, greenish. Legs, long, green. Summer – grey with black blotches above, head, breast and belly, heavily streaked black. Winter – pale grey above, neck and breast grey, very finely streaked. Estuaries and saltmarshes.

Limosa limosa
Black-tailed Godwit

(*North Sea, Baltic Sea, English Channel, Atlantic.*) To 43cm in length. Bill, long, almost straight, black tip, reddish base. Legs, long and black. Summer – head and breast, cinnamon-chestnut, upper parts dark brown and reddish mottle, tail white with black band seen in flight. Winter – uniform greyish above, underparts white. Estuaries and saltmarshes.

Limosa lapponica
Bar-tailed Godwit

(*North Sea, English Channel, Atlantic.*) To 40cm in length. Bill, long, upturned, black tip, pale orange base. Legs, shortish, black. Summer – head, neck and underparts red with dark streaks on crown and neck, upper parts dark brown and reddish mottle. Winter – grey above, highly streaked, white underparts. Mudflats and sandy shores.

Numenius phaeopus
Whimbrel

(*North Sea.*) To 43cm in length. Bill, long, downturned near tip. Legs, long and pale. Crown, striped white and dark brown, belly white, upper parts buff with indistinct darker markings. White rump in flight with barred tail. Mudflats and rocky shores.

Numenius arquata
Curlew

(*North Sea, Baltic Sea, English Channel, Atlantic.*) To 61cm in length. Bill, very long, evenly curved downwards. Legs, long, pale. Large, streaky brown, whitish chin and rump, barred tail in flight. Mudflats.

Birds

Recurvirostra avosetta
Avocet

(*North Sea, English Channel.*) To 45cm in length. Bill, long, very upturned, black. Legs, long, blue-grey. Crown and back of neck black, black bars on wings, remainder of body white. Estuaries, mudflats and brackish lagoons.

Ardea cinerea
Grey Heron

(*North Sea, Baltic Sea, English Channel, Atlantic.*) To 1m in length. Bill, heavy, sharply pointed, yellow, turning orange in spring. Legs, very long, brown, turning orange in spring. Blue-grey upper body, black crest on head. Head, neck and belly white. Saltmarshes and estuaries.

Phalacrocorax aristotelis
Shag

(*North Sea, English Channel, Atlantic.*) To 80cm in length. Bill, long, slender, small downward point, dark grey. Summer – glossy greenish-black, upcurved crest, wings with purplish hue, bright yellow gape patch. Winter – dull, no crest, smaller gape patch. Rocky shores.

Phalacrocorax carbo
Cormorant

(*North Sea, Baltic Sea, English Channel, Atlantic.*) To 1m in length. Bill, long, heavy, yellowish with small downturned point. Summer – head and body, glossy bluish-black, throat and cheeks white. Winter – dull, cheek and throat patches mottled brown. Rocky shores and cliffs.

Podiceps cristatus
Great-crested Grebe

(*North Sea, Baltic Sea, English Channel, Atlantic.*) To 51cm in length. Bill, long, sharply pointed, pink. Summer – crown black forming double crest with black and chestnut tippets. White face with black line from eye to bill. Brown back, white neck and belly. Winter – crown grey, reduced crest, no tippets, similar but more drab. Estuaries and open coasts.

Clangula hyemalis
Long-tailed Duck

(*North Sea, Baltic Sea, Atlantic.*) To 45cm in length. Bill, short, upturned. Male – head and neck white with grey patch round eye and darker area below, body white, grey flanks, blackish breast and wings, 13cm tail-streamers. Female – head blackish, buff sides and white ring round eyes, lower neck white, wing coverts and back brown, underparts white. All coasts.

Aythya fuligula
Tufted Duck

(*North Sea, Baltic Sea, English Channel, Atlantic.*) To 45cm in length. Bill, blue-grey. Male – glossy, purple head with drooping tuft, white belly and flanks, otherwise black. Female – dark brown, smaller tuft than male, speckled flanks. Open coasts and estuaries.

Aythya marila
Scaup

(*North Sea, Baltic Sea, English Channel, Atlantic.*) To 51cm in length. Bill, large and grey. Male – glossy green head, no tuft, white belly and flank, grey back, otherwise black. Female – brown, head dark with white face, flanks and belly pale and mottled. In eclipse both sexes duller. Coasts, especially bays and estuaries.

Anas penelope
Wigeon

(*North Sea, Baltic Sea, English Channel, Atlantic.*) To 48cm in length. Bill, small, grey. Male – head chestnut, forehead and crown yellow, body greyish with pink breast. Female – varying shades of brown, speckled. In eclipse, male and female like female but more speckled. Near muddy shores and eelgrass (p.130) beds.

Anas acuta
Pintail

(*North Sea, Baltic Sea, English Channel, Atlantic.*) To 59cm in length. Male with 8cm tail feathers. Bill, long, grey. Male – head chocolate, white below extending up head, back and flanks grey with black and cream markings. Female – pale, greyish-brown, dark crescents on flanks. In eclipse, male like female but darker above. Estuaries.

Anas platyrhynchos
Mallard

(*North Sea, Baltic Sea, English Channel, Atlantic.*) To 62cm in length. Bill, longish, yellow. Male – head, green, white neck ring, breast chestnut, pale back and belly, tail white with black, curled central feathers. Female – brownish buff with dark patterning. In eclipse, male like female. All coasts.

Anas crecca
Teal

(*North Sea, Baltic Sea, English Channel, Atlantic.*) To 38cm in length. Male – head, chestnut with green eye band, breast pale cream with brown mottling, body grey with white wing line and cream patch at tail. Female – brown with varying degrees and shades of speckling. In eclipse, male like female. Estuaries.

Mergus albellus
Smew

(*North Sea, Baltic Sea, English Channel.*) To 43cm in length. Male – mainly white, black on back, wings and between eye and bill. Female – cap chestnut, cheeks white, back dark grey, underparts light, speckled grey. Estuaries.

Mergus merganser
Goosander

(*North Sea, Baltic Sea, English Channel, Atlantic.*) To 69cm in length. Bill, long, pointed, red. Male – head greenish-black, back black, tail grey, remainder white. Female – head rufous with a crest, chin and breast white, remainder grey. In eclipse, both as female but slightly darker and duller. Occasionally estuaries.

Mergus serrator
Red-breasted Merganser

(*North Sea, Baltic Sea, English Channel, Atlantic.*) To 61cm in length. Bill, long, pointed, red. Male – head greenish-black with crest, white neck band, reddish breast with black speckling, black back, grey flanks. Female – head rufous with crest, white throat, rest mottled greyish-brown. In eclipse, male like female but blacker back and wings. Sheltered coasts.

Bucephala clangula
Goldeneye

(*North Sea, Baltic Sea, English Channel, Atlantic.*) To 48cm in length. Bill, short, grey. Male – head, glossy green with white blob between eye and bill, back and tail black, underparts and breast white. Female – head chestnut with white collar, back dark grey, underparts light grey. In eclipse, male like female but browner and no white collar. Coasts and estuaries.

123

Birds

Somateria mollissima
Eider

(*North Sea, Baltic Sea, English Channel, Atlantic.*) To 61cm in length. Bill, wedge-shaped, greyish-yellow. Male – white, with black crown, flanks, belly and tail. Pale green patches on back of neck, and breast tinged pink. Female – brown, speckled and streaked dark brown and blackish. In eclipse, male becomes black or dark brown, female darker. All coasts.

Tadorna tadorna
Shelduck

(*North Sea, Baltic Sea, English Channel, Atlantic.*) To 64cm in length. Sexes similar, female more drab. Mainly white with green head, chestnut body band, largely black wings, green secondaries, chestnut tertiaries. Male – bill bright red with knob. Female – dull red, no knob. Estuaries, sandy and muddy shores.

Melanitta nigra
Common Scoter

(*North Sea, Baltic Sea, English Channel, Atlantic.*) To 51cm in length. Male – glossy black, bill yellow with black knob. Female – almost black crown, greyish-white cheeks and throat, remainder dark brown, bill greenish-black with small knob. Sheltered coasts.

Branta leucopsis
Barnacle Goose

(*North Sea, Atlantic.*) To 69cm in length. Sexes similar. Long neck and small bill. Face creamy with black line from eye to bill, crown, neck and breast black, upperparts steely-grey, underparts white with pale grey barring. Mudflats, saltmarshes and machair.

Branta bernicla
Brent Goose

(*North Sea, English Channel, Atlantic.*) To 61cm in length. Sexes similar. Stocky with short neck. Face, neck and breast black with whitish neck markings. Upperparts grey-black and underparts only slightly lighter with dark-bellied race; alternatively, upperparts grey-brown and underparts pale. Mudflats and saltmarshes, particularly near eelgrass beds.

Anser anser
Greylag Goose

(*North Sea, Baltic Sea, Atlantic.*) To 89cm in length. Sexes similar. Large bird. Shades of brown. Head, neck and breast buff, longitudinal brown lines on neck, upperparts brown with pale bars, underparts grey with dark markings. Legs pink, bill orange. Saltmarshes.

Anser albifrons
White-fronted Goose

(*North Sea, English Channel, Atlantic.*) To 76cm in length. Sexes similar. Slight build. Face white with black border, head and neck dark brown, upperparts greyish-brown with pale bars, underparts darkish-brown with black bars, pale breast. Orange legs. Greenland race with orange bill, Russian race with pink bill. Saltmarshes, estuaries.

Anser brachyrhynchus
Pink-footed Goose

(*North Sea, English Channel, Atlantic.*) To 76cm in length. Sexes similar. Roundish head with short neck. Head and neck dark brown, upperparts pinkish-grey with whitish bars, underparts pinkish-brown. Legs pink, bill pink with black tip. Sandy shores, saltmarshes and estuaries.

Fulmarus glacialis
Fulmar

(*North Sea, English Channel, Atlantic.*) To 50cm in length. Head, neck and underparts white, upperparts grey. Stout, tube-nosed bill. Constant chuckling when sitting on nests on cliffs. Often seen gliding, skimming the wave tops. All coasts.

Stercorarius skua
Great Skua

(*North Sea, English Channel, Atlantic.*) To 61cm in length. Stocky, short tail and stout bill with hooked tip. Upperparts dark brown, underparts paler brown. Brilliant white flash on wings. All coasts.

Stercorarius parasiticus
Arctic Skua

(*North Sea, Northern Atlantic.*) To 48cm in length. Wings, narrow and pointed, tail with central feathers long and tapering. Plumage variable, from brown overall to dark brown above and almost white underparts.

Rissa tridactyla
Kittiwake

(*North Sea, English Channel, Atlantic.*) To 43cm in length. Small, slim head. Wings, slender and pointed. Head, underparts and tail white, back and upper surface of wings dark grey with white trailing edge and black tips. Bill greenish-yellow, legs grey. Breeds on cliffs in spring, otherwise oceanic.

126

Larus argentatus
Herring Gull

(*North Sea, Baltic Sea, English Channel, Atlantic.*) To 59cm in length. Stocky with short, broad wings. Head, neck, breast, tail and underparts white, back and wings grey. Wings have white fringe and black tips. Bill, yellow with red spot, legs pink. Complex array of juvenile plumages. All coasts.

Larus canus
Common Gull

(*North Sea, Baltic Sea, English Channel, Atlantic.*) To 43cm in length. Similar markings to Herring Gull, but smaller, longer winged, rounder head and delicate bill. Bill and legs yellow. All coasts.

Larus marinus
Great Black-backed Gull

(*North Sea, Baltic Sea, English Channel, Atlantic.*) To 69cm in length. Short, broad wings. Black back and wings, white line along trailing edge and wing tips. Body white. Bill, yellow with red dot, legs flesh coloured. All coasts.

Larus fuscus
Lesser Black-backed Gull

(*North Sea, Baltic Sea, English Channel, Atlantic.*) To 56cm in length. Back and wings slate grey, blacker towards tips, white dot near tip and white trailing edge. Body white. Bill, yellow with red dot, legs yellow. All coasts.

Larus ridibundus
Black-headed Gull

(*North Sea, Baltic Sea, English Channel, Atlantic.*) To 38cm in length. Slim. Rounded head, pointed wings. Summer – blackish-brown head. Winter – white head, black mark behind eye. Back and wings pale grey with white wedge, black tipped trailing edge. Body white. Bill and legs, dark red. All coasts.

Sterna hirundo
Common Tern

(*North Sea, Baltic Sea, English Channel, Atlantic.*) To 36cm in length. Slender bird with deeply forked tail, shorter or equal in length to closed wings. Head with black cap, upperparts, blue-grey, underparts white. Bill, red with black tip, legs red. All coasts.

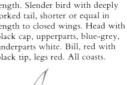

Sterna paradisaea
Arctic Tern

(*North Sea, Baltic Sea, English Channel, Atlantic.*) To 39cm in length. Slender bird with deeply forked tail, slightly longer in length than closed wings. Head with black cap, upperparts blue-grey, underparts greyish. Bill and legs red. All coasts.

Sterna albifrons
Little Tern

(*North Sea, Baltic Sea, English Channel, Atlantic.*) To 26cm in length. Small bird, narrow wings, short, forked tail. Head with black cap extending down back of neck and eye stripe, upperparts blue-grey, underparts and forehead white. Bill, yellow with black tip, legs orange. Sandy and shingle shores.

Sterna sandvicensis
Sandwich Tern

(*North Sea, English Channel, Atlantic.*) To 43cm in length. Large, long thin wings, forked tail. Head with black cap and crest, upperparts whitish-grey with darker wing tips, underparts white. Bill, slender, black with yellow tip, legs black. Rocky, shingle and sandy shores.

Alca torda

Razorbill

(*North Sea, Baltic Sea, English Channel, Atlantic.*) To 43cm in length. Pointed tail, short legs, large feet. Upperparts black, throat and cheeks chocolate in summer, white in winter, white line from eye to bill, underparts white. Black bill with white line. Cliffs.

Uria aalge

Guillemot

(*North Sea, Baltic Sea, English Channel, Atlantic.*) To 44cm in length. Rounded tail, short legs, large feet, slender pointed bill. Upperparts black, underparts white. Summer – head and neck chocolate. Winter – crown black, throat and cheeks white. Black bill. Inshore waters, cliffs.

Cepphus grylle

Black Guillemot

(*North Sea, Baltic Sea, English Channel, Atlantic.*) To 35cm in length. Smallish, short bill. Body, black-brown, white patch on wings, white underwings. More white in winter. Black bill with red gape, red legs. Rocky coasts.

Fratercula arctica

Puffin

(*North Sea, English Channel, Atlantic.*) To 31cm in length. Large head, large colourful bill, short tail. Upperparts black, underparts and cheeks whitish-grey. Red legs. Winter – colours more drab and greyer white, legs yellow. Inshore waters, cliffs.

Sula bassana

Gannet

(*North Sea, English Channel, Atlantic.*) To 96cm in length. Head and neck, creamy yellow, wing tips black, otherwise bright white. Bill, long and pointed, grey, legs dark grey. All coasts.

Flowering Plants (Angiospermae)

Only a very few flowering plants have managed to adapt to surviving in the marine environment, and those are in the genus *Zostera*. However, above the high water mark there is a considerable number of species that flourish, despite occasional immersion by the sea; these are the plants of the saltmarshes and coastal fringe. In the splash-zone, there are even more species that are adapted to living under the influence of the sea, but not being immersed by it.

Identification of flowering plants is dependant on various features, not all of which are always evident; the size and shape of the overall plant, flower characteristics, leaf shape and arrangement, degree of hairiness, colour and the habitat, may all be relevant.

Zostera marina

Eelgrass

(*North Sea, Baltic Sea, English Channel, Atlantic.*) Leaves to 50cm in length, 10mm wide, blunt or pointed end, with four or more parallel veins. Flowers green, insignificant. Lower shore.

Zostera angustifolia

Narrow Leaved Eelgrass

(*North Sea, Baltic Sea, English Channel, Atlantic.*) Leaves to 30cm in length, 2mm wide with notched end, and one to three veins. Flowers green, insignificant. Middle shore downwards.

Atriplex littoralis

Grass-leaved Orache

(*North Sea, Baltic Sea, English Channel, Atlantic.*) To 1m in height. Leaves narrow, shallowly toothed, mealy. Seven to eight small, greenish flowers.

Rumex crispus

Curled Dock

(*North Sea, Baltic Sea, English Channel, Atlantic.*) To 1m in height. Leaves to 30cm in length, narrow with curled edges. Seven to nine flowers, dense, green turning to red.

Halimione portulacoides

Sea Purslane

(*North Sea, Baltic Sea, English Channel, Atlantic.*) To 80cm in height. Shrubby. Leaves elliptical, grey, mealy, lower ones opposite, upper ones alternate. Seven to nine small, yellowish-orange flowers.

Atriplex prostata

Spear-leaved Orache

(*North Sea, Baltic Sea, English Channel, Atlantic.*) To 80cm in height. Leaves triangular, very slightly toothed, mealy below. Seven to nine small, greenish flowers.

Suaeda maritima

Annual Seablite

(*North Sea, Baltic Sea, English Channel, Atlantic.*) 50cm in height, prostrate to erect. Succulent, hairless annual, glaucous or red tinged. Leaves narrow, pointed, tapering to base, alternate. Eight to ten small, green flowers. Usually found below high water mark.

131

Beta maritima

Sea Beet

(*North Sea, Baltic Sea, English Channel, Atlantic.*) To 1m in height. Sprawling, succulent, hairless, often reddish stems and leaves. Leaves variable size and shape, often basal rosette, dark green, untoothed. Seven to nine green flowers.

Salicornia europea

Glasswort

(*North Sea, Baltic Sea, English Channel, Atlantic.*) To 20cm in height. Stems and branches succulent. Green, reddening at tips. Leaves scale-like, translucent. Eight to nine green flowers.

Salsola kali

Prickly Saltwort

(*North Sea, Baltic Sea, English Channel, Atlantic.*) To 60cm in height. Typically prostrate, prickly annual. Stem much branched, green with red stripes. Leaves narrow, succulent, sharp tipped. Seven to ten green flowers.

Honkenya peploides

Sea Sandwort

(*North Sea, Baltic Sea, English Channel, Atlantic.*) To 15cm in height. Creeping perennial with greenish-yellow succulent stems and oval leaves. Five to eight flowers, greenish white, to 10mm.

Spergularia marina

Lesser Sea Spurrey

(*North Sea, Baltic Sea, English Channel, Atlantic.*) To 20cm in height. Creeping annual, rather fleshy. Leaves succulent, linear, variably pointed. Six to eight flowers, usually pink, to 8mm.

Sagina maritima

Sea Pearlwort

(*North Sea, Baltic Sea, English Channel, Atlantic.*) To 15cm in height. Spreading from central rosette, fleshy, dark green. Leaves linear and blunt. Five to nine, small white flowers.

Silene maritima

Sea Campion

(*North Sea, English Channel, Atlantic.*) To 25cm in height. Erect to almost prostrate. Leaves variable in size, roughly pointed oval in shape. Six to eight flowers, solitary or in pairs, white with inflated bladder at base, to 2.5cm.

Crambe maritima

Sea Kale

(*North Sea, Baltic Sea, English Channel, Atlantic.*) To 60cm in height. Stout, fleshy, branched stems. Glabrous and basal leaves ovate, long-stalked, crinkly lobes. Upper leaves narrow. Six to eight flowers, white, to 10mm.

Lepidium latifolium

Dittander

(*North Sea, Baltic Sea, English Channel, Atlantic.*) To 1.3m in height. Much branched, glabrous stems. Long-stalked, ovate basal leaves with finely toothed margins. Flowers on six to seven whitish, pyramidal inflorescences, to 2.5mm.

Matthiola sinuata

Sea Stock

(*North Sea, English Channel, Atlantic.*) To 60cm in height. Many shoots, very leafy at base, hairy. Basal leaves stalked, lanceolate, lobed. Stem leaves narrowly elliptical, stalked. Six to eight pale pink flowers, to 2.5cm.

133

Cakile maritima

Sea Rocket

(*North Sea, Baltic Sea, English Channel, Atlantic.*) To 45cm in height. Erect or prostrate, glabrous, succulent, greyish. Lower leaves with stalk-like base, entire or lobed. Upper leaves not stalked. Six to eight flowers, lilac or white, to 10mm.

Cochlearia officinalis

Common Scurvy Grass

(*North Sea, Baltic Sea, English Channel, Atlantic.*) To 50cm in height. Prostrate, tending to rise at end, hairless. Basal leaves heart-shaped, long stalked and fleshy. Upper leaves clasping stem. Five to eight white flowers, to 10mm.

Raphanus maritimus

Sea Radish

(*North Sea, English Channel, Atlantic.*) To 80cm in height. Simple or branched stem, bristly especially above. Leaves pinnate with large terminal lobe, stalked, dark green. Six to eight yellow flowers. Characteristic seed pod.

Sedum anglicum

English Stonecrop

(*North Sea, Baltic Sea, English Channel, Atlantic.*) To 4cm in height. Mat-forming, glaucous, evergreen often red tinged. Leaves small, globular. Six to nine flowers, white above, pink below, to 12mm.

Vicia lutea

Yellow Vetch

(*North Sea, English Channel, Atlantic.*) To 60cm in height. Prostrate, shaggy, glabrous. Leaf with three to ten pairs of leaflets ending in tendrils. Six to eight pale dirty yellow, solitary flowers. Pods usually downy.

Lathyrus japonicus
Sea Pea

(*North Sea, Baltic Sea, English Channel, Atlantic.*) To 90cm in height. Creeping and ascending, glaucous, glabrous. Stem angled. Leaf, three to four pairs elliptical leaflets, end in tendrils. Six to eight flowers, purple to blue, to 22mm.

Eryngium maritimum
Sea Holly

(*North Sea, Baltic Sea, English Channel, Atlantic.*) To 60cm in height. Stem, intensely glaucous. Leaves leathery, spiny. Seven to eight flowers, pale blue, with surrounding spiny, bracts.

Crithmum maritimum
Rock Samphire

(*North Sea, English Channel, Atlantic.*) To 45cm in height. Branched stems, longitudinally ridged. Leaves pinnate with fleshy lobes, enfolding stem. Six to eight flowers, many in umbel, yellow.

Apium graveolens
Wild Celery

(*North Sea, Baltic Sea, English Channel, Atlantic.*) To 1m in height. Erect, grooved stem, strong smelling. Lower leaves simply pinnate. Upper leaves tri-lobed. Six to eight flowers, many in terminal and axial umbels, greenish–white.

Glaux maritima
Sea Milkwort

(*North Sea, Baltic Sea, English Channel, Atlantic.*) To 30cm in height. Creeping, curving upwards, glabrous. Leaves elliptical-oblong, succulent, almost unstalked, opposite. Six to eight flowers, pale pink, tiny, to 4mm.

Armeria maritima

Thrift, Sea Pink

(*North Sea, Baltic Sea, English Channel, Atlantic.*) To 30cm in height. Low growing, cushion forming. Leaves narrow, linear, with single vein, fleshy, glabrous, to 15cm. Six to eight flowers in roundish head on erect, leafless stem, various shades of pink.

Limonium vulgare

Sea Lavender

(*North Sea, Baltic Sea, English Channel, Atlantic.*) To 30cm in height. Branched, stout, rounded stems. Leaves variable in shape, broadly elliptical to lanceolate, with pinnate veins, long stalked. Seven to ten flowers, tightly packed, flat-topped clusters, lilac-lavender.

Limonium binervosum

Rock Sea-lavender

(*North Sea, English Channel, Atlantic.*) To 20cm in height. Branched stem. Numerous leaves, oval to lanceolate, not pinnately veined, tri-winged stalk. Seven to nine tightly packed flowers, curved spikes, lilac-lavender.

Mertensia maritima

Oyster Plant

(*North Sea, Baltic Sea, English Channel, Atlantic.*) To 60cm in height. Prostrate, straggling, mat-forming, glabrous, succulent. Leaves oval, fleshy, dotted with glands, tasting of oysters. Six to eight flowers, pink turning blue-purple, in clusters, to 6mm.

Plantago coronopus

Buckshorn Plantain

(*North Sea, Baltic Sea, English Channel, Atlantic.*) To 10cm in height. Leaves variable, nearly entire, toothed or most frequently pinnately lobed. Five to seven flower inflorescences, to 4cm, yellowish-brown on a 5cm reddish stalk.

Plantago maritima

Sea Plantain

(*North Sea, Baltic Sea, English Channel, Atlantic.*) To 15cm in height. Leaves in tuft, narrowly linear, fleshy, entire or only slightly toothed, faintly veined. Six to eight flower inflorescences, yellowish-brown on greenish stalk, to 6cm.

Matricaria maritima

Sea Mayweed

(*North Sea, Baltic Sea, English Channel, Atlantic.*) To 50cm in height. Branching stem, prostrate, turning up at end. Leaves oblong in outline, pinnate and feathery, whole plant almost glabrous. Seven to nine flowers, yellow centre, white rays.

Aster tripolium

Sea Aster

(*North Sea, Baltic Sea, English Channel, Atlantic.*) To 1m in height. Stout, erect stem, glabrous. Leaves fleshy, glabrous, faintly tri-veined, maybe slightly toothed, narrowly oblong. Seven to ten flowers, yellow centre, pale purple or whitish rays.

137

Inula crithmoides

Golden Samphire

(*North Sea, English Channel, Atlantic.*) To 60cm in height. Fleshy, glabrous stem. Leaves glabrous, linear, narrowing to base, maybe with three teeth at apex. Seven to eight flower heads, yellow rayed in loose cluster.

Artemisia maritima

Sea Wormwood

(*North Sea, Baltic Sea, English Channel, Atlantic.*) To 50cm in height. Very aromatic, branched, greyish, downy. Leaves bipinnate, ultimate segments linear, often white woolly hairs all over. Eight to nine flower heads, drooping, rayless, yellow, to 2mm.

Scilla verna

Spring Squill

(*North Sea, English Channel, Atlantic.*) To 20cm in height. Bulb, leaves narrow, grass-like, glabrous appearing before flowers. Four to five flowers on stem, flowerhead of two to twelve flowers, sky blue, rarely white, to 1.5cm.

Scilla autumnalis

Autumn Squill

(*North Sea, English Channel, Atlantic.*) To 20cm in height. Bulb, leaves narrow, grass-like, almost glabrous, appearing after flowers. Seven to nine flowers on stem, flowerhead of four to twenty flowers, purplish-blue, to 1.2cm.

GLOSSARY

Abdomen   The hindmost part of an arthropod's body
Algae   Simple plants including all seaweeds
Alluvial   Material transported and deposited by water
Alternate branching   Branches arising singly and on alternate sides of the stem
Antenna   Long slender projection at the head end of some worms and arthropods
Appendage   Projection from body, usually used for walking or feeding
Articulated   Movable
Biocide   Chemical that is toxic or lethal to life
Bract   Small leaf-like structure
Byssus   Hair-like filament used by mussels to attach to rock or a plant
Calcareous   Made of calcium carbonate
Carapace   External covering of some arthropods (e.g. crabs)
Chaeta   Bristle on an annelid
Cirrus   Small appendage on an arthropod or polychaete
Crenulate   Edge cut into small segments
Dichotomous   Divided into two parts
Digitate   Consisting of finger-like projections
Dorsal   Upper surface of a bilaterally symmetrical animal
Entire edge   Undivided, smooth edge
Epiphyte   An animal or plant growing attached to a plant
Filter-feeder   An aquatic animal that feeds by extracting suspended particles from the water
Frond   Leaf-like part of a seaweed
Ganglion   Solid mass of nervous tissue
Girdle   Muscular tissue that surrounds the shell valves of a chiton
Glabrous   Hairless
Glaucous   Greyish-blue colouration
Holdfast   Attachment organ in a seaweed
In eclipse   Bird in non-breeding plumage
Invertebrate   An animal without a backbone
Lamina   A thin layer
Lanceolate   Shaped like the head of a lance
Larva   A phase in the life cycle of an animal or plant, which is usually unlike the adult
Lateral line   A line of sense organs running down the side of a fish
Longitudinal   Running lengthwise
Machair   Low-lying sandy/boggy pasture
Mantle   The part of the body wall of a mollusc that secretes the shell
Opposite branching   Branches arising in pairs on opposite sides of the stem
Osculum   Opening in a sponge through which water is taken in
Ostium   Opening in a sponge through which water is expelled
Papilla   Small outgrowing structure on surface of an animal
Parapodium   Paddle-like appendage, often with chaetae or bristles, on a polychaete
Photosynthesis   The process by which plants use light energy to produce food
Phylum   The first level of division used when classifying organisms
Pinnate   Having a number of regular division or lobes
Polyp   An individual in a colony of animals
Proboscis   A protruding structure at the front end of some animals
Rhizoid   Root-like structure
Rostrum   Pointed structure at front end of the head of a crustacea
Runoff   Discharge from a river, stream or drain
Spatulate   Shaped like a spoon
Spicules   Minute, pointed structures, forming part of the skeleton of some animals (e.g. echinoderms)
Stipe   The stalk of a seaweed which connects the holdfast to the frond

Stylet	A small, pointed, bristle-like structure
Test	The 'shell' of a sea urchin or starfish
Thallus	The body of a seaweed or lichen
Thorax	The middle part of the body of an arthropod between the head and abdomen
Tranverse	At right angles to the long axis
Tripinnate	Divided into three
Tubercles	Small rounded projections
Umbel	Umbrella-shaped flower head
Umbilicalis	The opening in the central pillar of a snail shell
Umbo	Part of a bivalve shell
Valve	The individual shells of chitons
Ventral	Lower surface of a bilaterally symmetrical animal

Further Reading

Sea Life of Britain and Ireland, edited by Elizabeth Wood, published by Immel, 1988

Seas and Islands, The Living Countryside, consultant Keith Hiscock, published by Reader's Digest, 1984

Secrets of the Seashore, The Living Countryside, consultant Derek Hall, published by Reader's Digest, 1984

The Ecology of Rocky Shores, by J.R. Lewis, published by Hodder and Stoughton, 1976

Index

Index

Index